D0087383

The Great
Ocean Conveyor

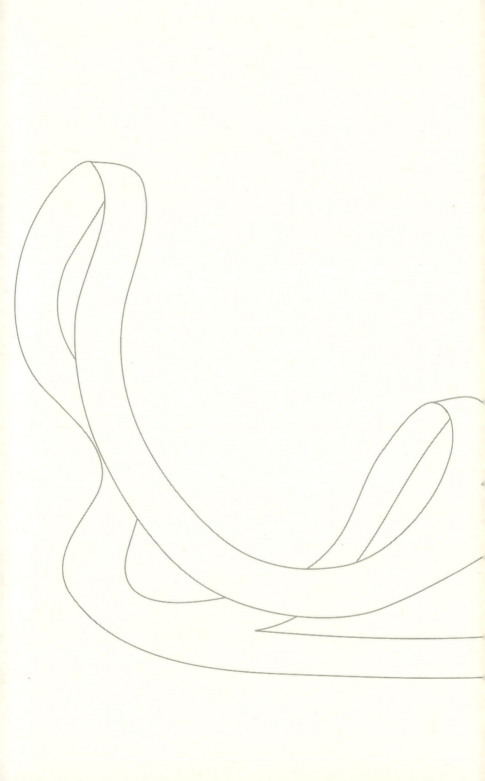

WALLY BROECKER

The Great
Ocean Conveyor

Discovering the Trigger
for Abrupt Climate Change

Princeton University Press
Princeton and Oxford

Copyright © 2010 by Princeton University Press

Published by Princeton University Press, 41 William
Street, Princeton, New Jersey 08540
In the United Kingdom: Princeton University Press, 6
Oxford Street, Woodstock, Oxfordshire OX20 1TW

Library of Congress Cataloging-in-Publication Data

Broecker, Wallace S., 1931–
The great ocean conveyor : discovering the trigger for
abrupt climate change / Wally Broecker.
p. cm.
Includes index.
ISBN 978-0-691-14354-5 (hardcover : alk. paper)
1. Ocean-atmosphere interaction. 2. Atmospheric
circulation. 3. Ocean circulation. 4. Climatic changes.
5. Boundary layer (Meteorology) I. Title.
GC190.2.B76 2010
551.5'246—dc22   2009030877

British Library Cataloging-in-Publication Data
is available

This book has been composed in Minion Pro

Printed on acid-free paper. ∞
press.princeton.edu
Printed in the United States of America
1 3 5 7 9 10 8 6 4 2

Contents

Preface

As a result of the concern over the impacts of the ongoing buildup of fossil fuel CO_2, studies of past climate have intensified. The hope is that information gained from these endeavors will help us better prepare for what is to come. Although our primary guide to the future will remain the simulations carried out in coupled atmosphere-ocean computer models, they have, as yet, proven incapable of replicating some important features of the paleo record. The reason is that they fail to properly represent powerful amplifiers and feedback mechanisms present in the real-world system; thus the interplay between these two ways of looking at the climate system has become an important aspect of our science.

I was fortunate to be the first one to propose that the Earth's ocean-atmosphere system has more than one stable mode of operation. In a sense, it is quantized. I realized this when I came up with an explanation for the abrupt coolings recorded in Greenland ice. I proposed that they had to do with the disruptions of a key element of the ocean's operation, namely, the meridional overturning circulation that takes place in the Atlantic Ocean. When northward-flowing warm upper-ocean water reaches the northern end of this ocean, it is cooled by frigid winter winds. This densifies it to the point where it sinks to the abyss and flows

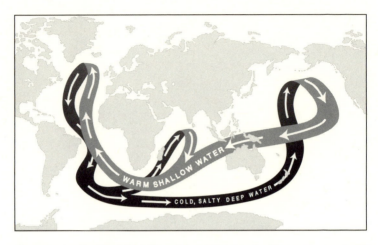

Figure 1. An idealized depiction of the path taken by deep water produced in the northern Atlantic Ocean, as prepared for a 1987 article in *The Natural History Magazine*.

southward, following the idealized loop shown in the figure. In an article published in *The Natural History Magazine*, I dubbed this loop "The Great Ocean Conveyor." My idea was that the abrupt coolings seen in Greenland were caused by sudden shutdowns of this current. At least one person strongly objected to my simplified diagram because she was greatly annoyed that the conveyor's lower limb passed straight through her New Zealand homeland!

This book recounts not only the circumstances surrounding my discovery but also how in its aftermath my thinking has evolved. Rather than following a linear path, progress has been rather chaotic, involving false leads and blind alleys. Hence, the flip flops I refer to later in this book are not only those of the conveyor but also those which have taken place in my head. Perhaps I should be embarrassed by the number of times I've had to change my mind, but I'm not. This is the nature of studies of our Earth's history. It is as if we view things through a dense fog. Whereas the primary object (in my case, the conveyor) is visible, all the secondary ones are pretty much obscured. So we make what we consider to be intelligent guesses as to what they are. A few such guesses prove to be correct but most turn out to be wrong. Should the guesses have been avoided? I don't think so, because our attempts to check whether or not we've got the guesses right lead to new discoveries that put us back on the correct track.

Put in another way, it's really almost miraculous that we know anything about past climate. There were no reliable thermometers or rain gauges until the eighteenth century, and before that our knowledge is based on indirect indicators for the real things of interest. We call these stand-ins "proxies." I will explain some of these later, but as none of them gives a clear or truly unambiguous record of what we really want to know, they must all be interpreted using analogues to the present or using models of some sort. Sometimes it turns out that these interpretations are inadequate because they fail to account for some unsuspected process. And sometimes it turns out that the observations themselves are just plain misleading: maybe not even indicating what we thought at all. Very little in science is known with certainty, and we must always be on guard against these problems. It is therefore not surprising that there have been some wrong turns, some misunderstandings, and some disputes. I have tried to be honest about these. Sometimes the attempt to explain some erroneous "facts" has even led to new ideas that have proved useful later on. That is what real science is like: a continual struggle to understand more fully and more accurately how the world really works.

Unlike mathematics or theoretical physics, which are often carried out in a single brain, our science is a communal one. We depend heavily on what others learn and on interactions with colleagues. As a dyslexic, I receive my most valuable information and ideas from what I hear and diagrams I see rather than what I read on the printed page.

In this book, I recount my efforts over the last twenty or so years to understand the message the paleoclimate record carries with regard to our climate system's capacity to undergo abrupt changes in its mode of operation. Hence, it must not be viewed as a comprehensive review of the subject. Rather, I recount only those observations and ideas that have significantly influenced my thinking. Therefore, I apologize to those whose efforts did not make a deep enough dent in my mind.

During my fifty-six years as a scientist, a host of people have helped me along. In hopes of not offending those not acknowledged here, I will mention only five who have made particularly large impacts on my thinking with regard to abrupt climate change.

A couple of years after my discovery of the conveyor, I used some money provided to Lamont-Doherty Earth Observatory by Exxon Corporation to conduct a three-week field trip to view the late Quaternary glaciations of the southern Andes. Each of the twenty-four participants was assigned a roommate. I chose my best friend, George Kukla, but after one night of enduring his awesome snores I switched to George Denton, someone I had met at meetings over the years but had never gotten to know very well. This switch changed both of our lives and it led to a scientific alliance that is now in its twentieth year. I, a man of the ocean, and he, a man of the mountains, realized that together we could make a difference. And we certainly have. I will never forget our first attempt. We invited ourselves to a meeting of the elite paleoceanography group that was held in Edinburgh, Scotland. Our mission was to convince these Milankovitch (see chapter 1) aficionados of the importance of millennial climate change. We bombed. On our flight back to the States, I tried to console a morose George by telling him that these things take time. I told him that in five years everyone

Figure 2. George Denton, University of Maine

in paleoclimate would be wild about abrupt change and Milankovitch's orbital cycles would have been relegated to the back burner. That's pretty much what happened.

In 1986, I spent a semester sabbatical leave at Caltech. Together with Barclay Kamb, I organized a seminar on the Earth's orbital cycles. One of the small group of attendees was Larry Edwards, who was finishing his PhD research on a new method for measuring ^{230}Th by mass spectrometry. I went back to Lamont and Larry took a teaching job at the University of Minnesota. Our only interaction, which came a couple of years later, involved a misunderstanding as to who should make measurements on some pristine glacial-age corals Rick Fairbanks had obtained offshore from the island of Barbados. Then a decade later, we reconnected and I became extremely interested in his research on stalagmites. It has had a profound impact on my thinking. Currently, we are working together on a project aimed at understanding how the Earth's rain belts responded to glacial conditions. Interactions with Larry energize me!

Figure 3. Larry Edwards, University of Minnesota

In 1992, I was asked by John Allen, the guru of the Biosphere 2 group that built the magnificent sealed greenhouse in the Arizona desert, for advice as to why the O_2 content of its atmosphere was declining and thus putting its eight occupants at risk. Jeff Severinghaus, who had just joined me as a graduate student, took up the challenge. The problem was that although it appeared that the microbes living in Biosphere 2's soil were eating O_2 faster than it was being produced by plants, the expected rise in CO_2 was not occurring. After some false starts and some sage advice from his father, Jeff showed that the missing CO_2 was being sucked up by the structural concrete that separated the ground floor, where the plants were grown, from the basement, which housed the mechanical equipment.

Jeff went on to do his thesis on the interaction among the gases (and their isotopes) in the sand dunes located in southernmost California. The important result that came out of this study was that Jeff discovered, by chance, that gases become fractionated

Figure 4. Jeff Severinghaus, Scripps Institution of Oceanography,
University of California

by temperature gradients in porous media. Armed with this interesting concept, he began, as a postdoc, what became his avocation, that is, exploiting the information contained in bubbles trapped at the base of the porous firn (transition zone from fluffy snow to solid ice) that caps polar ice. Now a professor at Scripps Institution of Oceanography, he has achieved fame by squeezing an amazing amount of information out of separations among these gases and their isotopes that were created while they were resident in the eighty-meter-thick dunelike firn. The results of these studies have added an enormous amount to what we know about abrupt climate change. I'm proud to have helped launch Jeff on his illustrious career and greatly value my continued interactions with him.

One day I received a telephone call from Charlie Bentley, a former Lamont PhD who is now at the University of Wisconsin. He sang the praises of Richard Alley, one of his graduate

Figure 5. Richard Alley, Pennsylvania State University

students, who was finishing his thesis on fabric of polar ice. I checked it out and concluded that Charlie was correct in his assessment. Richard would be a stellar catch for our faculty. But as we had just used up all our Brownie points in the hiring of Peter Schlosser, it was the wrong time to make the case for yet another faculty appointment. So Richard went to Penn State, where he remains despite our later attempts to lure him here to Lamont. Although not nearby, Richard has become a close friend and confidant. He serves as my "answer man." No matter what problem I pose, he promptly comes up with a solution. I was proud to be able to introduce Richard as the American Geophysical Union's Roger Revelle Medalist in December 2007, and in April 2008 I was pleased to learn that he had been voted into the National Academy of Sciences. Charlie's early assessment was certainly correct!

In April 2002, I received a letter from a man named Gary Comer. The letterhead read Lands' End Clothing. Not being

Figure 6. Gary Comer, founder of Lands' End, philanthropist

much of a consumer, I had never heard of the company or, of course, of Mr. Comer. In the letter he described an ice-free trip in his aluminum-hulled yacht, *Turmoil*, through the Northwest Passage. He sought advice regarding the state of the Arctic's sea ice from someone versed in the role of the ocean in climate change. In hindsight, it was my good fortune that he happened on my name. His visit to Lamont two weeks later launched what turned out to be a strong friendship involving a number of exciting joint ventures. Over the course of the next three years, I chose and he funded twenty-four world-class researchers in the area of abrupt climate change. We conducted field trips to Canada and to Greenland. We helped to launch a company that has developed a device capable of economically capturing CO_2 from the atmosphere. Finally, I sit at my desk in the fabulous new Geochemistry Laboratory financed by Gary Comer and named in his honor. Unfortunately, our friendship was cut short by his untimely death, a victim of cancer. But in those wonderful three years, he managed to reinvigorate me and send me off on several missions that even now keep me hopping.

So I dedicate this book to these five people who have made such big impacts on my quest to understand abrupt climate change.

The reader will note that I include very little material on modeling. One of the reviewers considered this to be a serious deficiency. The reason for this omission is that, with a couple of important exceptions, my thinking has been only marginally influenced by these computer-based simulations. In my estimation, they have had little predictive success. Rather, for the most part, they have played catch-up by attempting to duplicate paleoclimate observations. Two modeling efforts have rung my bell, however. Suki Manabe's ocean circulation model suggested that, while under interglacial conditions the ocean appeared to prefer a conveyor-on mode of operation, under glacial conditions it appeared to prefer an alternative conveyor-off mode. John Chiang's atmospheric model neatly explained the tropical impacts of conveyor shutdowns by showing that a freeze-over of the northern Atlantic would push the thermal equator southward, thereby significantly changing the pattern of rainfall. This is not to say that attempts to properly model what has gone on in the past are unimportant. They are, in fact, vital! Until models can satisfactorily reproduce the past, they will remain suspect with regard to telling us what the impacts of fossil fuel CO_2 will be.

Science is a fast-moving enterprise. It is characterized by a constant input of new information. As is illustrated here, concepts evolve. So take note that the last additions to this book were made in January 2009.

The Great
Ocean Conveyor

CHAPTER 1

The Setting

It was not until the mid-1980s that scientists became aware that our planet's climate system was capable of taking abrupt jumps from one state of operation to another. These jumps are the subject of this book. Before introducing them, however, we need to explore their context, namely, the stately progression of glaciations and interglaciations that are paced by cyclic changes in the configuration of the Earth's orbit (figure 1-1a).

Although, early on, physicists identified the precessing (that is, the slow gyration) and wobbling of our planet's spin axis as the likely drivers of the ice ages, geologists dragged their feet. In order to convince them that orbital changes were indeed the cause, during the 1920s and 1930s Milutin Milankovitch, a Serbian mathematician, made elaborate calculations elucidating the time sequence of seasonal changes in the amount of sunlight (i.e., solar insolation) reaching high northern latitudes. He reasoned that ice caps in North America and Europe likely grew during times of reduced summer insolation (that is, delivery of solar radiation) and retreated during times of enhanced summer insolation. In so doing, he hoped to provide geologists with a chronology to be compared with that for past glaciations. The problem was that, prior to World War II, geologic age

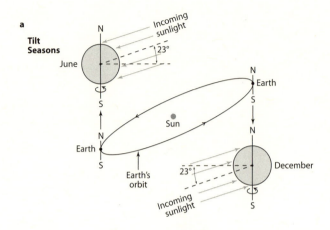

a

Tilt Seasons

Distance Seasons

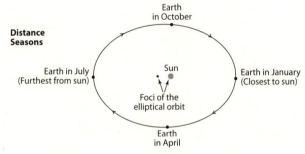

b

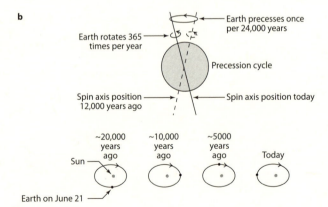

determination was in its infancy and hence incapable of providing the needed test. Thus, despite his painstaking calculations, Milankovitch failed to turn the heads of many of the day's paleoclimate specialists.

It was the postwar research of two professors at the University of Chicago that set the stage for the widespread acceptance of what had become known as the Milankovitch theory. Harold Urey, a chemist, demonstrated that the ratio of heavy oxygen (^{18}O) to light oxygen (^{16}O) in the calcium carbonate ($CaCO_3$) of seashells could be used to reconstruct past temperatures. Willard Libby, a physicist, harnessed a radioactive isotope of carbon, ^{14}C, to determine the ages of shell material formed during the last forty thousand years. Temperature and time, the two pieces of information required to determine whether or not Earth's climate has been paced by Milankovitch's orbital cycles, could now be determined.

What was lacking was a geologist with a vision regarding how to apply Urey's paleotemperature method to this problem. This vision was soon supplied by a young Italian, Cesare Emiliani, who, circa 1950, arrived in Urey's lab as a postdoc. He

---

Figure 1-1

a. The two sources of seasonality: that resulting from the *tilt* of the Earth's spin axis with respect to its orbit and that resulting from changing Earth–Sun *distance*. Cyclic changes in obliquity lead to a 41,000-year period in the amplitude of tilt seasonality. Cyclic changes in the roundness (i.e. eccentricity of the Earth's orbit lead to a 100,000-year period in the amplitude of the distance seasonality.

b. The precession of the Earth's spin axis leads to an antiphasing between distance seasonality experienced by the Northern Hemisphere and that experienced by the Southern Hemisphere. Switches occur with a period averaging about 20,000 years. The period is a bit shorter than the 24,000-year precession time because of the 105,000-year rotation of the Earth's orbit.

proposed to use Urey's mass spectrometer to conduct precise oxygen isotope analyses on tiny shells of surface ocean-dwelling (i.e., planktic) foraminifera abundant in deep-sea sediments. In the parlance of scientists, the resulting article, published in the *Journal of Geology* in 1955, constituted not only a home run but one with the bases loaded.

Rather than the brief cold snaps (glaciations) separated by long periods of warmth (interglaciations) envisioned by the paleontologist David Ericson based on his record of the relative abundance of species of planktic species, the oxygen isotope results indicated an alternation of warm and cold episodes of roughly equal duration. Furthermore, the duration of these episodes based on the extrapolation of sediment accumulation rates determined using Libby's ^{14}C method yielded a reasonable match to those predicted by Milankovitch.

While Emiliani awakened the paleoclimate community to the likelihood that Milankovitch had it right, one aspect of his interpretation of his results came under fire. He concluded that the glacial to interglacial cycles were accompanied by large (~7°C) swings in tropical ocean temperature. Emiliani was aware that the isotope composition of the oxygen in his shells was influenced not only by the temperature of the water in which the foraminifera grew but also that the waxing and waning of the ice sheets changed the isotopic composition of the ocean water in which they grew. The snow falling on the Greenland and Antarctic ice caps has an ^{18}O to ^{16}O ratio several percent lower than that of seawater. Hence, the growth of ice sheets must have led to an enrichment of ^{18}O in the sea. So, both cooling and ice growth contributed to the higher ^{18}O to ^{16}O ratio in foraminifera shells formed during times of glaciation. The problem was to estimate their relative contributions to the ^{18}O enrichment. Emiliani concluded that temperature dominated over ice volume.

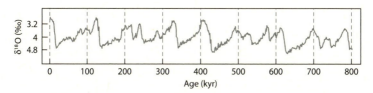

Figure 1-2. Benthic foraminifera ^{18}O to ^{16}O record for the past 800,000 years as obtained by Maureen Raymo of Boston University.

This conclusion was challenged by Nicolas Shackleton, a graduate student at Cambridge University whose great uncle, the explorer Ernest Shackleton, had been marooned in Antarctica's Weddell Sea when his wooden ship was crushed in the ice. Nicolas sought to check Emiliani's interpretation by doing oxygen isotope analyses on the bottom-dwelling shells of foraminifera (i.e., benthics). Shackleton's reasoning was as follows. As the temperature of deep seawater is already close to the freezing point, the contribution of temperature to the glacial enrichment of ^{18}O would have to be quite small. The problem he faced was the scarcity of benthics. It was not possible to get enough to provide the amount of sample required by the mass spectrometers of the time. So, Sir Nick,[1] as we now refer to him, set about to improve his mass spectrometer so that it could make accurate measurements on much smaller samples. It took several years before he succeeded. But the effort paid off. The oxygen isotope record for benthics had the same glacial to interglacial amplitude as Emiliani had found for planktics (see figure 1-2). This led Shackleton to conclude that ice volume rather than temperature dominated. After many years and many arguments, the

---

[1] Nicolas Shackleton stayed on at Cambridge and established himself as the dean of Cenozoic marine stratigraphy. His incredible list of accomplishments earned him not only important awards and membership in the Royal Society, but also a knighthood. Hence, "Sir Nick."

situation has finally settled down to a 60–40 split. About 1.05 per mil of the 1.75 per mil range in ^{18}O to ^{16}O ratio in the foraminifera records is the result of ice volume and about 0.70 per mil is the result of cooling. Hence, glacial-age tropical surface water cooled by only about 2.5°C instead of the whopping 7°C estimated by Emiliani.

Another contentious point was the age of the termination of the penultimate glaciation. Because the early ^{14}C measurements provided a reliable chronology back only about 30,000 years, some other chronometer was needed to reach further back in time. The obvious choice was ^{230}Th, an isotope with a 75,000-year half-life produced by the decay of uranium. Early on, this isotope was used to directly date deep-sea sediments. The approach took advantage of the fact that the ^{230}Th produced by the radioactive decay of uranium[2] dissolved in seawater was quickly absorbed onto particles and carried to the sea floor, providing newly deposited sediments with a large excess of this isotope. The radiodecay of this excess ^{230}Th provided the basis for sediment dating. This application was plagued, however, by the requirement that some means had to be adopted to take into account variations with time of the concentration of ^{230}Th in newly formed sediment. Two approaches were used. One, adopted by scientists at Columbia University, yielded an age of about 125,000 years for the end of the penultimate glaciation. Another, adopted by scientists at the University of Miami, confirmed Emiliani's original estimate of about 100,000 years.

---

[2] The ^{230}Th is produced by the decay of ^{234}U, which is a 254-kyr half-life daughter product of ^{238}U. It came as a surprise when a Russian geochemist discovered that ^{234}U and ^{238}U were separated when rocks are weathered. The reason is that, during their formation by alpha particle decay, the ^{234}U atoms are knocked loose and hence are preferentially released to solution. Seawater has a 15 percent excess of ^{234}U over that expected were it at steady state with its parent ^{238}U.

The argument over which of these approaches gave the best answer was an important one, for the 125,000-year age yielded a much better match to the Milankovitch summer insolation record. Once again, when the dust finally settled, Emiliani ended up on the wrong side of the argument. Although these two issues took a bit of shine off his efforts, the publication of Emiliani's 1955 article certainly marked the turning point in thinking regarding the pacing of the ice ages. He made it clear that Milankovitch had it right!

Even though Emiliani's article turned the tide, dissenters intensified their volley of criticism, as is often the case when major paradigm shifts occur. Further evidence in support of Milankovitch was needed! By serendipity it came when Robley Matthews, a geologist at Brown University, learned that the isotope ^{230}Th could be used to date corals. This new method was, in a sense, the inverse of that used to date marine sediments. Because ^{230}Th is efficiently stripped from seawater,[3] next to none is available to be built into corals. On the other hand, corals incorporate uranium as if it were calcium (i.e., the U to Ca ratio in corals is close to that in seawater). Then as coral ages, the ^{230}Th produced by the decay of uranium within the coral accumulates like the sand in the bottom of an hourglass. This buildup of ^{230}Th serves as a clock. Matthews asked David Thurber, a graduate student at Columbia University, to analyze two corals he had collected from raised terraces on the island of Barbados. His goal was to determine the rate at which the pore space in corals was filled in by diagenesis (something petroleum geologists were interested in, for ancient coral reefs constitute one of their favorite reservoirs).[4] When Thurber finished the analyses, there was great excitement

---

[3] The element thorium has a very strong tendency to attach itself to the water column particulates. Hence geochemists refer to it as a particle-reactive element.

[4] *Diagenesis* refers to chemical alteration of sediments as a result of mineral-pore fluid interactions.

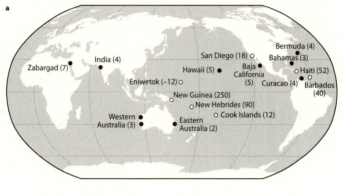

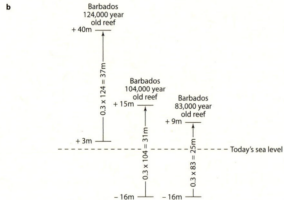

Figure 1-3

a. Locales where the 124,000-year old interglacial high sea stand has been radiometrically dated by the ^{230}Th-^{234}U method. The numbers represent their present-day elevations in meters. The solid circles represent tectonically stable coastlines, and the open circles from shorelines undergoing tectonic uplift.

b. Elevations of the three raised coral reefs on the island of Barbados. Based on the 3-meter height of the last interglacial sea stand (as determined on stable coastlines), the average uplift rate of Barbados has been about 0.3 meters per kyr. Based on this rate, 105,000 years ago and 83,000 years ago the sea stood about 16 meters below its present level. Keep in mind that at the peak of the last glacial period it stood about 120 meters lower than it does today.

in our lab because the age of one coral came out close to 124,000 years[5] and the other close to 83,000 years. Both were times when Milankovitch's calculations yielded maxima in Northern Hemisphere summer insolation and, accordingly, glaciers should have melted back, raising sea level (see figure 1-3). When asked about the setting of his coral terraces, Matthews explained that Barbados was being underthrust by the Atlantic's oceanic crust. As a consequence, it was being uplifted, pushing above sea level coral reefs formed during times when the sea stood lower than at present. Matthews then surprised Thurber by mentioning that there was a third coral terrace situated between the other two (i.e., below the elevation of the 125-kyr coral terrace and above that of the 83-kyr coral terrace). Thurber was eager to determine its age. The result turned out to be about 105,000 years. At first this was puzzling because the Milankovitch reconstruction had no peak in summer insolation at that time. This puzzle was resolved when it was realized that the Milankovitch reconstruction placed too much emphasis on the contribution of the 40,000-year obliquity cycle (i.e., the rocking of the Earth's orbit that alternately strengthened and weakened the tilt seasonality) and too small a contribution of the ~20,000-year precession cycle. Calculations based on a different choice of latitude than that chosen by Milankovitch led to a greater contribution of precession and produced a summer insolation peak at 105,000 years ago. Hence, Thurber's finding that the sea level reached three successive maxima in concert with three summer insolation peaks spaced at 20,000-year intervals provided confirmation that Milankovitch cycles paced fluctuations in the size of the Earth's ice sheets.

An even stronger verification based on an analysis of the spectral makeup of benthic foraminifera oxygen isotope records

---

[5] This age turned out to mark the onset of the last interglacial interval and was key to the resolution of the argument over the timing of this event.

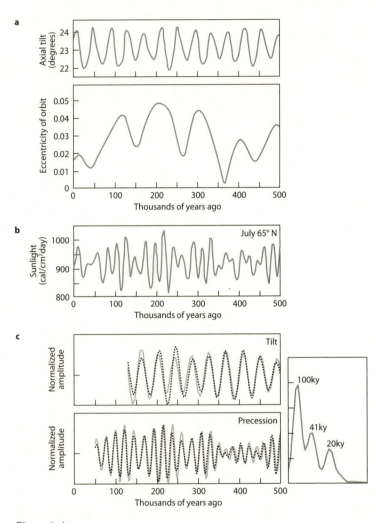

Figure 1-4

a. The Earth's axial tilt and orbital eccentricity over the past 500,000 years as determined by celestial mechanics.

b. July solar irradiance at 65°N over the past 500,000 years.

c. Amplitude of the 41,000- and 20,000-year spectral components of the benthic foraminifera $^{18}O/^{16}O$ record (dashed curve) compared with that of the tilt (i.e., obliquity) and distance (i.e., eccentricity–precession) came some years later. This effort, spearheaded by John Imbrie, a paleontologist at Brown University, showed that three periodicities dominated the benthic record, one close to 20,000 years ago, one close to 40,000 years, and one close to 100,000 years (figure 1-4). The first two of these were, as expected, reflecting the 20,000-year precession and the 40,000-year tilt cycles. The finding of the third periodicity (i.e., 100,000-year), however, created a puzzle that even today has not been satisfactorily resolved. Over the better part of the past million years, the oxygen isotope record for benthic foraminifera has undergone a cycle with an asymmetrical triangular shape. Long intervals of cooling accompanied by growing ice sheets terminated during relatively short time intervals, returning climate to its full interglacial condition. In other words, as the glacial world warms, the ice sheets melted away.

When examined in detail, however, the spacing between these terminations was never exactly 100,000 years. Rather, it was either close to 80,000 or close to 120,000 years. This gives the impression that during each glacial episode the Earth system drifted toward some sort of instability that, when reached, triggered a jump back to the warm state. Furthermore, these jumps occur preferentially during episodes of strong Northern Hemisphere summer insolation. But the exact nature of the instability that produces the termination remains a mystery.

Several kilometer-long cores drilled through the Antarctic and Greenland ice caps turn out to be the Rosetta stones of the

contributions to seasonality (solid curve). Due to "end" effects in the spectral analysis, the most recent several cycles had to be excluded. Also shown (to the right) is a so-called power-spectrum of a deep-sea ^{18}O record. This figure is reproduced from a paper authored by John Imbrie. This long-accepted evidence in support of insolation forcing has recently been challenged by Harvard's Peter Huybers. He shows that it is an artifact of the analysis procedure.

Ice Ages. The variety and detail of the information they contain is staggering. While the 800,000-year-duration record in Antarctic ice is dominated by the smooth cyclic variations orchestrated by the Milankovitch insolation cycles, that in Greenland is so riddled with the impacts of millennial-duration events that the Milankovitch imprint is largely masked. Furthermore, the record in Greenland covers only the past 130,000 years.

Building on the success of an American ice core from Byrd Station, Antarctica, Russian scientists decided to attempt to get a much longer record by drilling on the frigid polar plateau. Lacking transport aircraft, their scientists were forced to remain at the site year-round. During one of the frigid winters, their generator broke down and they survived by huddling in an ice cave heated only by candles. Despite this and other incredible hardships, after five long years they prevailed and penetrated 90 percent of the way through the ice cap. But, alas, they had no way to get the hundreds of tons of ice back to their laboratories. So it remained unanalyzed, stacked like logs at the frigid drill site.

Claude Lorius, a French scientist, realizing the immense value of this ice archive, negotiated an amazing deal. The ice would be flown out of Antarctica on American C147 cargo planes equipped with skis designed for snow landings and takeoffs. Part of the ice would go to a stable-isotope laboratory in France. Thus the first record of south polar temperatures covering several 100,000-year cycles was obtained. It revealed a surprise. As the precession cycle is antiphased between the hemispheres, if Milankovitch cycles were pacing climate, it would be expected that the 20,000-year periodicity in southern temperatures would be antiphased with that in the benthic ^{18}O record, which reflects the size of the northern ice sheets. Contrary to expectation, both records follow the tune of summer insolation at high northern latitudes. The reason for this remains a mystery.

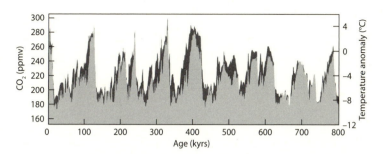

Figure 1-5. Record over the past 800,000 years from the Dome C Antarctica ice core of stable isotope–based temperature departures from today's (gray) and CO_2 contents of air trapped in bubbles (black) as obtained by French scientists.

In addition to a record of south polar temperatures, Antarctic ice contains a beautiful record of the changing trace gas composition of the Earth's atmosphere. Air bubbles trapped during the conversion of fluffy snow to solid ice constitute a pristine archive. Much information of great importance to our understanding of climate change is contained in this trapped air. Clearly, greatest interest is focused on the record of the atmosphere's CO_2 content. Measurements carried out in laboratories in both Bern, Switzerland, and Grenoble, France, showed that during times of peak glaciation, the CO_2 content of air was about 30 percent lower than during times of peak interglaciation. Furthermore, the shape of the CO_2 record is similar to that for oxygen isotopes in benthic foraminifera, faithfully following the stately Milankovitch pacing (but only weakly participating in the millennial disruptions that will concern us in the chapters that follow). And, as CO_2 is an important greenhouse gas, these ups and downs certainly contributed to the temperature changes that accompanied glacial cycles. Subsequent drilling by the European (EPICA) group at the polar plateau Dome C site yielded ice dating back to 800,000 years ago (see figure 1-5). Two aspects of this record are of particular interest. The first is the amazing similarity in the shape of the CO_2 and temperature records. The CO_2 content of the atmosphere is closely tied to the air temperature over the ice cap. The second is that both follow the Northern Hemisphere's summer insolation.

As there is currently no radiometric means of directly dating ice, initially the chronologies proposed for the Antarctic record were based on one or the other of two strategies. One took advantage of the similarity between the ice core and benthic foraminifera oxygen isotope records, and adopted the marine chronology. The other was based on the present-day rate of snow accumulation at the core site coupled with assumptions regarding the dependence of this rate on temperatures as reconstructed from the stable-isotope record. Of course, account also had to be taken of the thinning that occurred as the ice flowed toward the edges of the cap. As these two approaches agreed reasonably well, initially scientists were satisfied. As time went on, however, it became evident that a more precise chronology was needed if small leads on lags with respect to other records were to be reliably established.

Michael Bender, a scientist at Princeton University, came up with a clever means of tying the ice chronology directly to that for local Milankovitch's summer insolation cycles. As the latter chronology is based on rigorous celestial mechanics (i.e., on Newton's laws of gravity), it is highly precise. Bender's discovery came about somewhat by serendipity. He set out to make very accurate measurements of the ratio of O_2 to N_2 in the trapped air. His idea was that he might be able to detect the very small decreases in atmospheric O_2 during glacial time expected as the result of the oxidation of organic matter as the cold conditions killed the boreal forests. But he was disappointed (and puzzled) to find that the variations were far larger than could be attributed to vegetation oxidation. Intrigued as to what else could be the cause, he persisted and constructed an O_2/N_2 record that extended back several hundred thousand years. He was amazed to find that the fluctuations correlated beautifully with the amount of Antarctic summer insolation. He reasoned that this correlation was related to the observation that the greater the solar heating, the larger the individual ice crystals formed as the snow recrystallized. Perhaps the size of the ice crystals influenced the geometry of the "walls" surrounding the air bubbles as they were closed off in the final stage of lithification (i.e., the conversion of snow to solid ice). As closure occurs at a depth of many tens of meters, the air in the newly formed bubbles is under pressure. Hence, some is forced out through the remaining tiny orifices. As O_2 and N_2 molecules differ in size, Bender hypothesized that they are differentially squeezed out of the bubbles. As the extent of this separation would likely depend on the size of the ice crystals, it would also depend on summer insolation, Bender was able to make the case that his O_2 to N_2 ratio results created a firm tie to the local insolation changes created by the wobble of the Earth's orbit and the precession of its spin axis! The fact that the chronology he created in this way agreed quite well with the previous ones put to rest any lingering doubts regarding the validity of the Antarctic timescale.

One more record must be mentioned before we turn to the millennial-duration blemishes that mar the smooth swings paced by the Earth's orbital cycles. It is a record of the strength of the Asian monsoons contained in the stalagmites from caves in China. As with the record in marine foraminifera shells, it is based on ^{18}O to ^{16}O measurements on $CaCO_3$ (in this case stalagmite calcite). The chronology for the record is obtained in the same way as that for corals. Uranium picked up as rainwater percolates through the soil that overlies the cave is built into the calcite. But, as thorium remains behind safely locked in the soil, little accompanies the uranium. The ages are calculated by the amount of ^{230}Th that is subsequently built into the stalagmite by

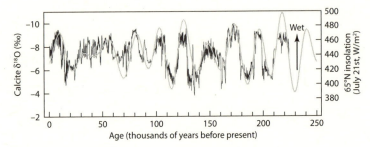

Figure 1-6. Strength of the East Asian monsoon as recorded by the [18]O to [16]O ratios in calcite from China's Hulu, Dongge, and Sanbao Caves. As shown by the smooth curve, the strength of the monsoon closely follows summer insolation at Northern Hemisphere high latitudes. The chronology was determined by precise [230]Th–[234]U dating at the University of Minnesota by Larry Edwards and Hai Cheng in cooperation with Chinese colleagues Wang Yongjin, Yuan Daoxian, and An Zhisheng, who kindly permitted them to be reproduced here.

the decay of uranium. As a graduate student at Caltech, Larry Edwards developed a mass spectrometric method to measure [230]Th (and also its parent, uranium). He demonstrated that in this way he could improve the measurement precision by an order of magnitude over that obtained by the decay-counting method used by Thurber to date Barbados corals. As a professor at the University of Minnesota, Edwards reduced the analytical uncertainty of his mass spectrometry method by yet another order of magnitude and demonstrated that it could produce fantastically precise ages. For example, he now is able to achieve an accuracy of ±60 years on samples with an age of 100,000 years. (By comparison, Thurber's errors by the decay-counting method were several thousand years!) The only other chronology with this level of accuracy is that for the seasonality of solar insolation, which is based on celestial mechanics. Working with Chinese postdoctoral fellows and graduate students, Edwards has been able to piece together results that together provide a continuous oxygen isotope record extending back more than 200,000 years (see figure 1-6). This record has been replicated in three caves. Although the same small temperature and ice volume–related ^{18}O changes recorded in foraminifera must be present, the ^{18}O to ^{16}O ratio variations in Chinese stalagmites are much larger. They are dominated by variations in the contribution of monsoon rainfall to the total annual rainfall.

The main feature of the stalagmite record is its remarkable resemblance to that of 65°N summer insolation. As the monsoons are driven by the summer heating of the Asian continent, this is not unexpected. What is surprising is that the 100,000-year cycle that dominates both the benthic foraminifera record and the Antarctic ice record is only weakly expressed. On the other hand, the millennial events, which are this book's main focus, show up beautifully as marked negative deviations from the summer insolation trend. We will, of course, have more to say about these deviations in the chapters that follow.

Before we leave this introduction, a few words are needed about the questions it raises. Clues to their answers will be found in information contained in the record of millennial fluctuations.

1) Why can't the world's most advanced atmosphere–ocean models reproduce the impacts of Milankovitch insolation cycles as seen in the paleoclimate record? As we shall learn from our consideration of the impacts of the abrupt millennial events, the system has powerful feedbacks related to the way the ocean circulates and to the presence of sea ice.

2) Why does the cycle of the Northern Hemisphere ice sheets have an asymmetric saw-tooth shape? Key to the answer of this question is the understanding of the abrupt terminations of each 100,000-year cycle. To date we lack this understanding.

3) Why does the temperature in Antarctica appear to follow the pacing of the Northern Hemisphere's orbital cycles? The

answer to this question likely lies in what we shall refer to as the ocean's bipolar seesaw.

4) What mechanism allowed the ocean to suck in and breathe out CO_2, thereby generating the glacial-interglacial cycle in the atmosphere's content? Despite many attempts over the past twenty-five years, no entirely satisfactory explanation has been forthcoming. But as we shall see, it very likely has to do with sea ice cover around Antarctica.

5) Why doesn't the strength of the Asian monsoons show a more pronounced influence of the asymmetrical saw-toothed cycle so evident in the other records? As we shall see, although the strength of the Asian monsoons is strongly affected by the presence of sea ice, it does not appear to follow the change in size of the ice caps on land. This suggests that heat released from the sea is critical!

CHAPTER 2

A Surprise

Willy Dansgaard, a Danish scientist long interested in the isotopic composition of rain and snow, was the first to demonstrate that the ratio of heavy oxygen (^{18}O) to light oxygen (^{16}O) in precipitation varied systematically with air temperature. For example, winter snow falling on Greenland's polar plateau contained about 3 percent (30 per mil) less heavy oxygen than did rainfall in the tropics. As has already been discussed, this ^{18}O deficiency in polar snow was responsible for the enrichment of ^{18}O in glacial ocean. Dansgaard further demonstrated that when the average isotopic composition of annual precipitation at high-latitude sites was plotted against each site's average annual temperature, the points fell nicely on a straight line. The ^{18}O to ^{16}O ratio decreased by 0.7 per mil for each degree Celsius cooling (see figure 2-1). He was pleased to find that this slope was close to that predicted based on the difference in vapor pressure between $H_2^{16}O$ (~1 percent) and $H_2^{18}O$ and the dependence of the saturation concentration of water vapor on air temperature (7 percent per degree). The idea was that, as an air mass was cooled along its poleward trajectory, water condensed and fell as rain or snow. Because of the vapor pressure difference between "heavy" and "light" water, the precipitation would have

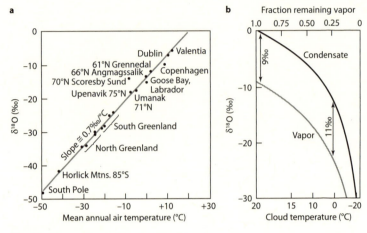

Figure 2-1.

a. Observed $\delta^{18}O$ in average annual precipitation as a function of average annual air temperature, compiled by Willi Dansgaard. Note that all the points on this graph are for latitudes greater than 45°. The $\delta^{18}O$ values are calculated as follows:

$$\delta^{18}O = \frac{^{18}O/^{16}Osample - \,^{18}O/\,^{16}Ostd.}{^{18}O/\,^{16}Ostd.} \times 1000$$

b. $\delta^{18}O$ in cloud vapor and condensate plotted as a function of the fraction of remaining vapor in the cloud. The temperature of the cloud is shown on the lower axis. The increase in isotope fractionation with decreasing temperature has been taken into account.

a slightly higher ^{18}O to ^{16}O ratio than the vapor from which it formed. Hence the vapor would become progressively depleted in ^{18}O (as would the precipitation formed from it).

With this in mind, Dansgaard was eager to use this relationship to determine how much colder Greenland was during the last glacial period. Realizing that the snow that fell on Greenland's interior flowed through the cap and eventually "calved" at the margins, forming icebergs, he sampled many of these in hopes of finding ice with even lower ^{18}O to ^{16}O ratios than occurred in snow currently accumulating on Greenland's high plateau. But as there was no way to determine the age of the ice in any particular berg, he soon realized that if he were to succeed, he had to get ice from a hole drilled through the ice in Greenland's interior.

The opportunity came when an American group did exactly this. Taking advantage of the logistics available at the air force base located at Thule in northwestern Greenland, during the late 1960s the Americans drilled through three kilometers of ice to bedrock at a site called "Camp Century." Dansgaard made detailed oxygen isotope measurements on the ice they recovered and showed that the record extended back through the entire last glacial period into part of the preceding interglacial period. He was pleased to see that the ice representing the peak of the last glacial period had, as expected, lower ^{18}O to ^{16}O ratios than ice from the Holocene. Using the slope of the present-day temperature-^{18}O relationship, he estimated that, on average, the glacial maximum interval was about 10°C colder than now. Only much later was it shown that this reconstruction gave too small a cooling. The actual one was closer to 23°C! The reason for Dansgaard's underestimate not only is quite interesting but has proven to be exceedingly informative. But this is getting ahead of our story.

Strangely enough, what proved to be the most important feature of this record was originally largely overlooked. It was not until fifteen years later, when a second long core was obtained in southern Greenland, that its significance came to light. Except for the intervals representing the glacial maximum and the Holocene, Greenland's oxygen isotope records are riddled with quasi-rectangular millennial-duration excursions (see figure 2-2). Each has an amplitude about half that of the glacial maximum to Holocene change. In the eyes of paleoclimatologists,

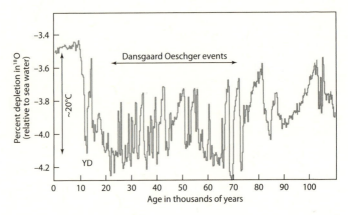

Figure 2-2. Unlike the stable isotope record in Antarctic ice, which is dominated by stately Milankovitch-paced cycles, that in Greenland ice is riddled with large and abrupt millennial-duration fluctuations known as Dansgaard-Oeschger events. The temperature scale on this diagram is based on a deconvolution of downhole thermal profile rather than on the ^{18}O measurements themselves, which suggest only half as large a temperature change as shown here. These measurements were made in the laboratory of Minze Stuiver at the University of Washington and in the laboratory of Willi Dansgaard in Copenhagen.

only the last of these excursions had a familiar look—that is, the outlier separating last glacial maximum ice from Holocene ice. Pollen records from Scandinavia had a century earlier revealed that the warm period (referred to as the Bølling Allerød) that marked the end of the last glacial period lasted for only about two millennia before giving way to a relapse into cold conditions. This cold episode was named the Younger Dryas, in recognition of a small alpine flower that as a result of this cold snap was able to migrate down to sea level. In Greenland ice, the Bølling Allerød–warm to Younger Dryas–cold (BA-YD) oscillation is beautifully recorded in the oxygen isotope record. What no one had ever seen before were the twenty or so

Younger Dryas–like ^{18}O events that punctuated that portion of the record predating the glacial maximum.

What is surprising is that when in 1972 Dansgaard first presented this record to an audience of climate scientists assembled at Yale, no one jumped up and shouted, "What the hell are those sharp changes?" In fact, his record was greeted with what could be described as yawns. Having witnessed this unveiling and having reacted in the same blasé manner as the other attendees, I have often wondered why there was no sense of amazement, especially since the Younger Dryas oscillation was not only well known but also a subject of considerable interest. I suspect that the reason was that we were, at that time, all caught up in the excitement created by Emiliani's marine ^{18}O record. Rather than marveling at what have become known as the Dansgaard-Oeschger events, we all puzzled as to why the twenty- and forty-thousand-year-duration Milankovitch cycles were not more evident.

A decade later another ice core was recovered, this time at the location of the Dye 3 radar base in southern Greenland. The ^{18}O record from this core looked much like that from Camp Century. Were it not for the accompanying CO_2 record, another decade might have passed before the significance of the millennial-duration events was recognized. But when, in the mid 1980s during a meeting in Switzerland, Hans Oeschger presented a detailed set of CO_2 measurements covering part of the time interval punctuated by the Dansgaard-Oeschger events, ears perked up (mine included), for accompanying each sudden rise in ^{18}O was a roughly 50 ppm rise in CO_2 concentration and accompanying each sudden drop in ^{18}O was a roughly 50 ppm drop in CO_2 concentration (see figure 2-3). As we geochemists were still struggling to explain the 90 ppm difference in atmospheric CO_2 content between full glacial and full interglacial time, the finding that 50 ppm changes had occurred in

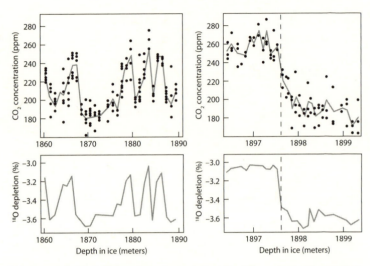

Figure 2-3. In the left-hand panel are plotted the CO_2 content of trapped air and the extent of ^{18}O depletion in the ice itself for a several-thousand-year time period centered at about thirty thousand years ago, as obtained by the Oeschger Group in Bern, Switzerland. During this interval, four Dansgaard-Oeschger events occurred. As can be seen in the figure, during the warm phase (lower ^{18}O depletion) of each Dansgaard-Oeschger event, the CO_2 content of the bubbles rises from about 190 ppm to about 240 ppm. These increases were brought into question when CO_2 results from Antarctica ice cores became available. All the measurements on air from Antarctic ice in this time interval were in the range of 190 to 200 ppm. The explanation appears to be that the high CO_2 values in the Greenland record were the result of the release within the ice of CO_2 generated as a result of a chemical reaction between $CaCO_3$ and acids contained in the ice. This explanation received strong support when it was shown by the Swiss group that the expected offset between the record kept in ice and that kept in air bubbles did not exist (see right-hand diagram).

just a few decades was mind boggling. I remember returning home from this meeting with all kinds of wild thoughts swimming through my mind. One thing was certain: any explanation had to involve moving CO_2 into and out of the ocean. As I had spent considerable time attempting to quantify the fraction of the CO_2 produced by fossil fuel burning that has been taken up by the ocean, I knew that making a 50 ppm change on a decade timescale posed a huge challenge.

My thinking was influenced by the requirement that not only the CO_2 change but also the air temperature change had to be explained. If changes in ocean circulation were responsible for the CO_2 change, they were also likely to have been responsible for the temperature changes. Furthermore, since the temperature change was observed in Greenland (and the BA-YD oscillation in Scandinavia as well) the place to look was at the northern Atlantic Ocean. The thought came to mind that perhaps it had to do with the formation of deep water in that region. During my graduate student days, I made measurements of radiocarbon on water samples raised from the deep Atlantic. The goal of these measurements was to determine the rate at which newly cooled water descending from the surface "ventilated" the deep Atlantic. Because of this, I was familiar with the influence of this process on the heat budget of the region surrounding the northern Atlantic. Waters warmed in the tropics were carried into the Norwegian Sea by an extension of the Gulf Stream. Frigid winter air flowing off Canada and Greenland extracted this heat, thereby cooling (and densifying) the water. This caused it to sink to join what oceanographers refer to as North Atlantic Deep Water mass.

I reasoned that the heat released during this process helped maintain northern Europe's moderate winters. This led to an idea that changed the way we think about paleoclimate. What if somehow the production of deep water were shut down? Wouldn't that lead to a dramatic cooling of the region around the northern Atlantic?

Although it was a possible explanation for the abrupt temperature changes seen in the ice core record, I could think of no means by which such a stoppage could significantly alter the atmosphere's CO_2 content. Also, my experience with the fossil fuel CO_2 budget made clear that pushing the required amount of CO_2 in and out of the ocean on such a short timescale was impossible.

Fortunately, the necessity to do this soon disappeared when it was demonstrated that rapid CO_2 excursions did not, after all, occur; rather they were shown to be artifacts of storage in Greenland ice. This retraction was forced by two sets of measurements made in Oeschger's laboratory. The first set covered the same time interval in an ice core from Antarctica. Instead of jumping back and forth between about 190 ppm and 240 ppm, as seen in Greenland's Dye 3 ice, all the measurements in the Byrd Station ice were close to 190 ppm. The second set of measurements was designed to determine whether or not the expected offset between the CO_2 record kept in air bubbles and the ^{18}O record kept in the solid ice was present. This offset occurs because bubble closure takes place at the base of the firn horizon many tens of meters below the surface. When a careful comparison between the ^{18}O and CO_2 profiles was conducted, the sharp change in CO_2 was shown to occur at exactly the same depth as the sharp change in ^{18}O. The only way to explain this was to postulate that the extra 50 ppm of CO_2 was added after the bubbles formed. As will be explained later in the chapter, this addition involved a reaction between $CaCO_3$ dust and airborne acids (HNO_3 and H_2SO_4) trapped in the ice.

So, it's rather bizarre that the observation that put me on the path that led to the Great Ocean Conveyor turned out to be a false alarm. But even though this idea had its twists and turns, it has withstood the test of time.

One of the many questions the conveyor hypothesis raised was what triggered the abrupt shutdowns of deep-water production and, of course, what triggered its abrupt restorations.

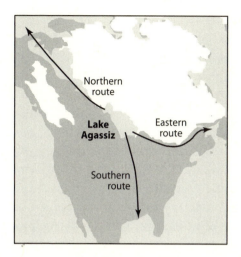

Figure 2-4. As the Laurentide ice sheet retreated at the close of the last glacial period, meltwater lakes formed around its margin. One of these, Lake Agassiz, underwent a major drop in level at about the time of the onset of the Younger Dryas cold snap. One hypothesis is that water released by this drop flooded to the east into the northern Atlantic, bringing the conveyor to a halt.

Although an entirely satisfactory explanation has yet to be proposed, early on a tantalizing explanation for the onset of the Younger Dryas caught my attention. As the Laurentide ice sheet retreated during the Bølling Allerød warm episode, a lake formed in the glacially downwarped terrain[1] along its southern margin (see figure 2-4). This lake is named in honor of Louis Agassiz, who first convinced the world that features long attributed to Noah's flood were instead produced by glaciers. Initially, the meltwater that filled Lake Agassiz spilled to the south over a rock lip and flowed down the Mississippi River into the Gulf

---

[1] The weight of the three-kilometer-thick ice cap caused hot mantle rock to flow laterally out from under the cap, lowering its surface by about one kilometer.

of Mexico. Then, as documented by both radiocarbon dating of Lake Agassiz's shorelines and of ^{18}O records for Gulf of Mexico sediments, something changed. According to the geologist Jim Teller at the University of Manitoba, a collapse occurred in the ice that formed the northern shoreline of the lake, thereby opening a new outlet. The outlet to the Mississippi was abandoned and the water instead flowed to the east through the basin of what is now Lake Superior and on to the Atlantic via the Gulf of St. Lawrence. The initial deluge released by the collapse of the ice barrier was just what was needed to squelch deepwater formation, for it would have reduced the salt content of the northern Atlantic surface waters to the point where winter cooling could no longer produce water dense enough to permit the cascade down into the abyss.

While this explanation provided the encouragement I needed to pursue the conveyor idea, even then doubts lurked. One was that although this deluge might have led to an abrupt shutdown, what was it that a thousand or so years later caused the abrupt restoration of deep-water formation? Another had to do with the multiple Dansgaard-Oeschger events. Could it be that each abrupt cooling was caused by a similar flood? Despite these questions, I stubbornly clung to this idea until in 2003 a group of geologists (accompanied by Gary Comer) conducted an aerial survey of the region to the west of Thunder Bay, Ontario, through which Teller claimed the Agassiz flood had passed. Not a shred of geomorphic evidence was found. No channels, no boulder fields. So again, one of the prime pieces of evidence that led me to propose the conveyor idea appears to have fallen by the wayside. But despite this, other evidence has kept this hypothesis alive.

During the mid-1980s, many of us took for granted that the climate impacts of the Younger Dryas were confined to the area surrounding the northern Atlantic. Although a few claims of impacts outside this region had been made, they were unconvincing and for the most part disregarded. As this distribution fit well with the concept that it was caused by the turning on and off of deep-water formation in the northern Atlantic, I used it as substantiating evidence that my idea was correct. But, like the CO_2 jumps and the Agassiz flood, this observation was subsequently proven to be incorrect. Among the evidence indicating that the distribution of Younger Dryas impacts extended well beyond the region around the northern Atlantic was that contained in Greenland ice. This evidence was recorded both in the impurities present in the ice and in the methane content of air trapped in the ice. Ice from the Younger Dryas and from the ultracold Dansgaard-Oeschger intervals has more impurities and less methane than that from the adjacent warmer intervals.

The impurities in ice are contributed by dust and by aerosols. Their concentrations are determined by melting the ice and measuring the ions contained in the water (see figure 2-5). The Ca^{++}, Mg^{++}, and K^+ were presumed to have been delivered as very fine dust, the Na^+ and Cl^- as sea spray, and the $SO_4^=$ and NO_3^- as aerosols created from the gases released from vegetation and swamps. Also present but not measured are the $CO_3^=$ ions from $CaCO_3$ dust. Of these impurities, only the provenance of the dust can be pinned down. Traces of the elements strontium, and neodymium are present. Each has an isotope produced by the decay of a long-lived radioactive parent. In the case of strontium it is ^{87}Sr, the decay product of ^{87}Rb, and in the case of neodymium it is ^{143}Nd, the decay product of ^{147}Sm. Analysis of dust from each of the world's deserts reveals that, taken together, the isotopic compositions of these two elements serve as a fingerprint.[2] Using this approach, Lamont-Doherty's Pierre Biscaye

---

[2] The reason is that the geologic processes responsible for local differences in ^{87}Sr are not identical to those responsible for local differences in ^{143}Nd.

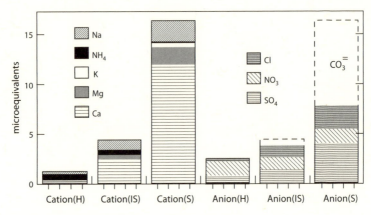

Figure 2-5. Cation and anion composition of Greenland ice during three climate states: the Holocene (H), Dansgaard-Oeschger interstadials (IS), and Dansgaard-Oeschger stadials (S). A charge imbalance exists because the $CO_3^=$ ions present as $CaCO_3$ were not measured. These analyses were performed in the laboratory of Paul Mayewski.

demonstrated that the dust in Greenland ice comes largely from the far-off Asian deserts. Hence conditions in this region must have been affected by the Younger Dryas!

One other measurement related to the impurities in the ice must be mentioned. It is the electrical conductivity. By dragging two sharp electrodes (with a difference of a thousand or so volts between them) across a clean ice surface, a measure of the concentration of hydrogen ion is obtained. The hydrogen ions present in the ice were emplaced by acid aerosols (i.e., mainly H_2SO_4 and HNO_3). Some of this acid is neutralized by reaction with $CaCO_3$ dust, however.

$$2HNO_3 + CaCO_3 \rightarrow Ca^{++} + 2NO_3^- + CO_2 + H_2O$$

$$H_2SO_4 + CaCO_3 \rightarrow Ca^{++} + SO_4^= + CO_2 + H_2O$$

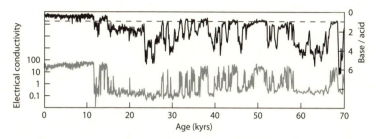

Figure 2-6. Electrical conductivity for the Greenland GISP2 ice core (obtained by Ken Taylor of Nevada's Desert Research Institute) for the last seventy thousand years plotted on a log scale. Also shown is the ratio of the contribution of acids (mainly HNO_3 and H_2SO_4) to that of bases (mainly $CaCO_3$).

Hence, in Holocene ice where the input of acids exceeds that of $CaCO_3$, the electrical conductivity is high, and in glacial ice where $CaCO_3$ exceeds acid, the conductivity drops to near zero (see figure 2-6).

The reason for the warm-to-cold ratio change is primarily the result of greater glacial dustiness. Examples of this are seen in the glacial-age loess (wind-blown dust) found in China, the United States, the Czech Republic, and Argentina. It is tempting to conclude that the reason for the high glacial dust flux is that the deserts were drier. But, as will be discussed later in the book, the closed basin lakes in the source regions for Greenland dust were far larger during glacial time. Hence these regions must have had more precipitation and hence were *less* wet! The explanation is more likely that during glacial time latitudinal temperature gradients were stronger and, as a result, the planet was stormier, causing more dust to be lofted high into the atmosphere.

Because it is so detailed, the electrical conductivity record speaks to the abruptness of transitions that bound the millennial events. They are completed on a timescale of a few decades. One might argue that the abruptness reflects the nonlinearity of conductivity rather than reality, but even when plotted on a logarithmic scale, the transitions remain very sharp.

We can now pick up a point left dangling early in the chapter, namely, why there was extra CO_2 in Greenland bubbles at the times of the Dansgaard-Oeschger moderate cold events but not in the corresponding times in Antarctica. As mentioned earlier, the excess carbon dioxide associated with Greenland's episodes of moderate cold must have been produced within the ice. Why isn't the same thing seen in Antarctica? The answer to this question is that the content of dust in Antarctic ice is more than an order of magnitude less than that in Greenland ice. Hence, there is not enough $CaCO_3$ in southern polar ice to generate measureable CO_2 anomalies. This raises another question. Why do the CO_2 results from only the moderately cold Dansgaard-Oeschger intervals in the Greenland ice disagree with those from Antarctica? For times of interglaciation, the answer is clear. The acid dissolves the $CaCO_3$ before the bubbles close, allowing the extra CO_2 to diffuse up through the firn and out into the overlying atmosphere. The problem comes in explaining why excess CO_2 was not produced in ice representing ultracold times. During these intervals there was more $CaCO_3$ than acid. One would have to argue that when $CaCO_3$ dominates, the acid was entirely neutralized in the firn before the bubbles closed. But we must explain why this did not happen when the acid and CO_2 were present in roughly equal amounts. The only thing I can think of is that perhaps the acids were preferentially deposited in summer ice and the $CaCO_3$ preferentially deposited in the winter ice. During the Holocene, there was enough acid in both summer and winter ice to get rid of the $CaCO_3$ before lithification was complete and during the peak glacial there was enough $CaCO_3$ in both to do in the acid. But during the moderately cold times, acid was preserved in the summer layers and $CaCO_3$ in the winter layers. Only well after lithification was it possible that some combination of layer thinning and molecular diffusion allowed them to react with each other to produce CO_2. As this is a bit difficult to understand, table 2-1 may help.

TABLE 2-1

Possible scenario designed to explain the excess CO_2 found in the interstadial horizons of Greenland ice

Interglacials	Acid destroys all $CaCO_3$ present in firn
	CO_2 formed in this way escapes from firn
	Residual acid accumulates in ice
D-O Stadials + Glacial	$CaCO_3$ neutralizes all acid present in firn
	CO_2 formed in this way escapes from firn
	Residual $CaCO_3$ accumulates in ice
D-O Interstadials	In summer layers, acid destroys all $CaCO_3$
	CO_2 formed in this way escapes from firn
	Residual acid accumulates in ice
	In winter layers, $CaCO_3$ neutralizes all acid
	CO_2 formed in this way escapes from firn
	Residual $CaCO_3$ accumulates in ice
	During storage in ice, summer layer residual acids react with winter layer residual $CaCO_3$, producing CO_2, which remains trapped in ice

Now back to the far-field impacts recorded in Greenland ice. A second line of evidence comes from the methane record stored in the air bubbles. The main source of methane is wetlands and fresh-water swamps. Because O_2 diffuses orders of magnitude slower through water than through air, water-logged sediments tend to be anaerobic. The demand for oxygen by decaying vegetation exceeds the supply of oxygen from the overlying atmosphere. In the absence of other abundant oxidants such as sulfate ions, bacteria cause organic molecules to react with one another to produce methane and carbon dioxide, which then diffuse out into the atmosphere. As the lifetime of methane molecules in today's atmosphere is close to a decade and has probably remained close to that despite changes in climate, the variations in the concentration of methane in the air trapped in ice provide a measure of the production rate of methane at any given time in the past. Hence the ice core record provides a measure of the extent of wetlands and swamps in the past.

What the record shows is that during the Younger Dryas, the methane content of the atmosphere underwent an abrupt decrease. As the swamps and wetlands responsible for supplying the atmosphere's methane during this time interval were largely in the tropics, the suggestion is that the impacts of the Younger Dryas extended into the tropics. By implication, the tropics were less wet during the Younger Dryas interval. More on this later.

Our next step will be to consider the records from other climate archives. As we shall see, they reinforce what has been learned from the Greenland ice cores: namely, the impacts of the millennial-duration events extended throughout the Northern Hemisphere and into the tropics. But before we consider these records, it makes sense to take a diversion and discuss what is known about the ocean's thermohaline circulation. The Great Ocean Conveyor is one branch of this vast system of currents.

CHAPTER 3

The Villain

As turning off and turning on the Great Ocean Conveyor is the basis for my idea regarding the cause of the abrupt changes seen in Greenland's ^{18}O to ^{16}O record, it is important to place this element of the ocean's circulation system into context. The ventilation of the ocean's interior can be subdivided into three parts.[1] Thermocline ventilation refers to the upper ocean. Thermohaline ventilation refers to the deep ocean. Separating these two realms are intermediate waters that penetrate each of the three oceans from the south. The Great Ocean Conveyor describes a branch of the thermohaline circulation involving deep waters formed in the northern Atlantic. They account for about half of the water supplied to the abyss over the last millennium. The other half was generated along the margin of the Antarctic continent.

We know from radiocarbon measurements that the average time a water parcel resides in the deep sea before being returned to the surface by upwelling is on the order of one thousand years. Upwelling matches sinking from the surface. In order

---

[1] *Ventilation* is a word used by oceanographers to denote the steady replacement of *in situ* water by that descending from the surface. Hence it is analogous to the replacement of room air with that from the outside.

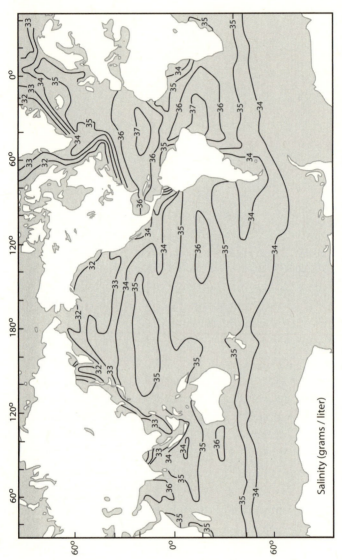

Salinity (grams / liter)

Figure 3-1. Map of the distribution of mean annual surface ocean salt content. Note that in each ocean a salinity maximum exists in the latitude range 15° to 30° (i.e., in regions where evaporation exceeds rainfall). Also note that salinities in the northern Pacific are everywhere about 2 g/liter lower than those at the same latitude in the northern Atlantic Ocean.

to sink into the deep sea, water must be a bit denser than the resident water. Seawater becomes denser as it is cooled. Also it is densified by salt. Hence new deep water forms in those areas on the globe that have the coldest and saltiest water. Hence the term *thermohaline circulation*. Close to seawater's freezing point (i.e., −1.8°C), a cooling of 1°C is equivalent to an increase in salt content of about 0.2 grams per liter. Deep waters in the ocean have salt contents falling within the range 34.6 to 35.0 grams per liter and temperatures in the range 0° to 3°C.

A quick look at the map of surface salinities (figure 3-1) eliminates one region of the polar ocean from contention as a source of new deep water. Salinities in the northern Pacific are very low, falling in the range 32 to 33 grams per liter. As a result even when cooled to their freezing point, they do not come close to being dense enough to sink into the deep sea.

This leaves only two contenders, the northern Atlantic and the margins of the Antarctic continent (i.e., the Southern Ocean). It turns out that deep water formed in the Southern Ocean descends at close to its freezing point with a salt content close to 34.3 g/liter. In the northern Atlantic, surface waters have a salt content a tad below 35 g/liter and when cooled to a temperature in the range of +1° to +3°C are dense enough to sink into the abyss. Because these waters are 3° to 5°C warmer than those sinking in the south, their extra 0.7 g/liter of salt is not quite enough to make them as dense as the deep water formed in the Southern Ocean. This being the case, when the two waters "collide," that formed in the northern Atlantic tends to override that formed in the Southern Ocean.

As depicted in figure 3-1, the anomalously low-salt-content water found in the northern Pacific extends southward into the tropics. At any given latitude, Pacific surface waters are 1 to 2 g/liter lower in salt content than their Atlantic equivalents. It is this difference that provides the impetus for deep-water formation in the Atlantic Ocean. Surprisingly, the excess in the Atlantic's salt content is the result of an interplay of the Earth's wind systems with its mountain ranges. This interaction leads to a net export of water vapor from the Atlantic Ocean to the Pacific Ocean. The salt left behind as a consequence of this water vapor export concentrates salt in the Atlantic and the water vapor gain by the Pacific dilutes its salt. The rate of this transfer is so large that were the salt gain by the Atlantic not compensated by a net export, its salt content would rise at a rate of roughly 1 g/liter each 1,500 years. As thermohaline circulation appears to have remained more or less constant during the course of the last 10,000 years (i.e., during the Holocene), this tendency toward buildup must have been continuously compensated by the export of salt to the Pacific. This departing salt is carried by the lower limb of the Great Ocean Conveyor. High-salinity deep waters generated in the northern Atlantic are transported by a complex system of currents and eddies down the length of the Atlantic and around the tip of Africa into the Southern Ocean, where they merge with newly formed Southern Ocean deep water and with deep water recirculated from the deep Pacific and Indian oceans. This mixture peals off from the circumpolar "mixing machine" and moves up the abyssal Pacific Ocean (and also up the abyssal Indian Ocean). In this way the excess salt is removed from the Atlantic and eventually reunited with the excess fresh water delivered as the result of vapor transport to the Pacific.

Maintenance of this balance between fresh-water transport through the atmosphere and salt transport through the sea is presumably accomplished by a self-regulatory mechanism. For example, were the Atlantic's excess salt not removed rapidly enough, its salt content would increase, speeding up the export from the Atlantic, thereby tending to eliminate the mismatch.

Before continuing the discussion of ocean's thermohaline circulation, a few words about vapor transport are in order. The

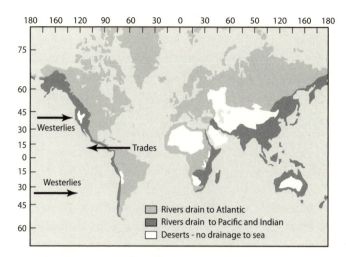

Figure 3-2. Map of the global drainage basins. The boundary between the gray and black regions corresponds to the divide separating drainage to the Atlantic from that to the Indian and Pacific Oceans. The white areas are deserts from which no river outflow occurs. Also shown are the directions of the major wind systems.

most important topographic barrier to vapor transport is the American Cordillera. These high mountains prevent vapor carried by the temperate latitude westerly-wind systems (see figure 3-2) from reaching the interior of the Americas, for in order to enter the continent, air masses must rise to a high elevation. The consequent cooling causes almost all of their water vapor to precipitate as rain on the western flanks of the mountains. It then runs back into the Pacific where it originated.

By contrast, the westerlies entering Europe (and Africa) do not encounter any such topographic barrier. And although they lose their moisture to rainfall, much of which runs down rivers that feed back into the Atlantic, a substantial amount re-evaporates and continues its transit across the continent, eventually reaching the Pacific. Hence the net influence of the westerly winds is to transport water vapor from the Atlantic to the Pacific.

Despite the American Cordillera, tropical easterly winds are able to carry large amounts of water vapor from the Atlantic to the Pacific. The reason is that the gap in this mountain chain that made possible the Panama Canal allows the trade winds to freely move across the narrow isthmus. The equatorial mountains in eastern Africa prevent the equivalent transport of Indian Ocean moisture across Africa into the Atlantic.

The establishment of the exact magnitude of the net water vapor transport from Atlantic to Pacific proves extremely difficult. Estimates have been made using a combination of wind and humidity data, using atmosphere general circulation models, and using water budgets (i.e., rainfall, evaporation, and river runoff). The results range from 0.15 to 0.35 million cubic meters per second. When compared to the discharge of the world's rivers (1 million cubic meters per second) or to the global rainfall (15 million cubic meters per second), this transport is small potatoes—hence the difficulty in narrowing the range in the estimates.

A portion (about 0.10 million cubic meters per second) of the salt enrichment due to vapor loss is compensated by the inflow to the Arctic of low-salinity northern Pacific waters (via the Bering Straits). The rest is compensated by exchanges across the South Atlantic–Southern Ocean interface (see figure 3-3). The outflow via the conveyor's lower-limb water is about 16 million cubic meters per second and it carries a salt excess of roughly a 0.25 g/liter. Hence it compensates for roughly (.25/34) × 16 or 0.12 million cubic meters per second of water vapor loss. Depending on their proportions, the combination of the inflow of salt-rich Indian Ocean surface water passing around the tip of Africa into the Atlantic and of the salt-poor water passing around the tip of South America into the Atlantic could either

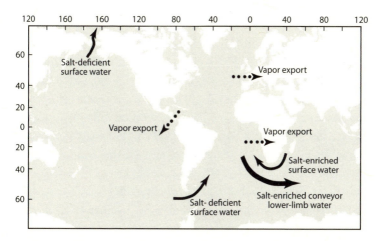

Figure 3-3. Map depicting the transports involved in the salt budget of the Atlantic Ocean.

increase or decrease the net removal of salt. The bottom line is that, although we know that the export of salt must balance the gain resulting from vapor loss, it is beyond our current capability to determine exactly how this balance is accomplished.

There are two ways by which the relative contributions of the northern Atlantic and Southern Ocean source water to any given sample of deep water might be estimated. The obvious approach would be to use temperature and salinity data. But as waters produced at different places in the northern Atlantic have ranges in both temperature and salinity, this approach results in sizable uncertainties. The alternative approach is to use a property known as PO_4^* (phosphate star), which is defined as follows:

$$PO_4^* = PO_4 + \frac{O_2}{175} - 1.95$$

Its utility depends on the observation that for each molecule of PO_4 released during respiration, 175 molecules of O_2 are consumed. Hence, despite the fact that the dissolved oxygen carried by newly formed deep water is gradually consumed and that the phosphate it carries is gradually enhanced, the PO_4^* value remains unchanged. Furthermore, the PO_4^* value for all "brands" of newly produced northern Atlantic deep water lies in the narrow range 0.73 ± 0.03. That produced in the Southern Ocean has a similarly narrow range and, importantly, a much larger value of 1.95 ± 0.05. Taken together, the small range for each end member and the big difference between the end members make PO_4^* a much better source indicator.

One might ask, why are these two PO_4^* values so different? The answer is that although the two source waters have similar initial O_2 contents, Southern Ocean water has a much higher initial phosphate content. It is similar to that for the mix of deep waters circulating around the Antarctic continent. Since this water upwells under ice and is rapidly cooled and reoxygenated before being sent back down as new deep water, there is little chance for it to lose its PO_4 to plant growth.

By contrast, the PO_4 content of deep water formed in the northern Atlantic is set by its content in the inflowing surface waters (which match the outflow via the conveyor's lower limb). Something like half of this water comes from the Indian Ocean around the tip of Africa into the South Atlantic. Phytoplankton inhabiting the Indian Ocean deplete this water's phosphate. Hence, even though the other two sources—water coming in around the tip of South America and water coming in from the northern Pacific via the Arctic—have moderately high phosphate contents, the low-phosphate Indian Ocean component holds down the average. Important to keep in mind in this regard is that, although the residence time of phosphorus in the ocean with respect to removal to sediments is several tens of millennia, the residence time of water in the Atlantic is only about two centuries. Hence the amount of phosphorus carried out of the Atlantic by the conveyor's lower limb *must* be exactly the same as that entering with the inflow waters. As a result, the phosphate content of deep water formed in the northern Atlantic is about half that in newly formed Southern Ocean water. This accounts for the large difference between PO_4^* values for deep water formed in the northern Atlantic and that formed in the Southern Ocean.

The distribution of PO_4^* values in the deep ocean tells us some very important things about how today's deep ocean currents operate. First, the PO_4^* values for all the waters deeper than 1,500 meters in the Pacific Ocean are very close to 1.34, a value that lies halfway between that for northern Atlantic source water (0.73) and that for Southern Ocean source water (1.95). This tells us that, at least when averaged over the residence time of water in the deep Pacific (~1,000 years), the contributions of the two sources have been about equal.

Second, the distribution of PO_4^* values in the deep Southern Ocean tells us that this mixture is created in a single pass around the Antarctic continent (see figure 3-4). For this reason I refer to the circumpolar region as the ocean's "mixing machine." Waters rich in the northern end member hug the African margin as they enter the Southern Ocean. Being less dense, they tend to override the ambient circumpolar water. Waters rich in the southern end member slide down the margin of the Antarctic continent into the Southern Ocean. Being more dense, they tend to underride ambient circumpolar water. These waters enter mainly in the Atlantic sector. As can be seen in figure 3-4, the contrast in PO_4^* created by these additions, while strong to the south of Africa, steadily diminishes along the path of the circumpolar current and by the time the water reaches the Drake Passage and reenters the Atlantic sector, a nearly uniform mixture has been created! Furthermore, this mixture has a PO_4^* value of 1.4 (i.e., close to that for Pacific deep water).

a

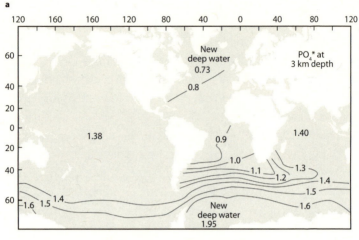

b

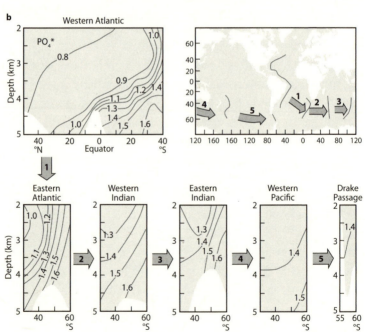

Third, although the North Atlantic Deep Water mass has a reasonably uniform PO_4^* value, as the result of mixing with circumpolar water, it is everywhere somewhat higher than that for the newly formed end member (0.73). Furthermore, its PO_4^* value increases toward the south. The North Atlantic Deep Water mass is sandwiched between two water masses with high PO_4^* values. The lower one is known to oceanographers as Antarctic Bottom Water and the upper one as Antarctic Intermediate Water. Both originate in the Southern Ocean and gradually lose their identity as they penetrate northward in the Atlantic.

Of course, if we are to understand the consequences of the turning off and on of the Great Ocean Conveyor, we must know how the ocean operated during glacial time. This is a tough challenge and, despite much effort by paleoceanographers, it has proven elusive.

What we do know comes from information stored in marine sediments, and the key element of this record is the carbon isotope ratio in the shells of planktic and benthic foraminifera. Although we have no direct proxy for paleophosphate, it turns out that the stable carbon isotope ratio of ^{13}C to ^{12}C provides a

---

Figure 3-4

a. The distribution of PO_4^* (i.e., $PO_4 + O_2/175 - 1.95$) at 3 kilometers depth in the world ocean. Deep waters formed in the northern Atlantic have PO_4^* values in the range 0.73 ± 0.03 and those formed in the Southern Ocean values in the range 1.95 ± 0.05. The two end members are homogenized in the circum-Antarctic ring current, creating a mixture consisting of about 50 percent northern component and about 50 percent southern component. This mixture floods the deep Pacific and Indian oceans.

b. Sections showing how PO_4^* changes down the western Atlantic and around the circum-Antarctic ring. As can be seen, the contrast between waters rich in the northern component and those rich in the southern component seen to the south of Africa is largely mixed out during a single pass around the Antarctic continent.

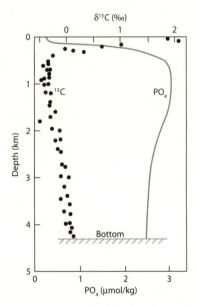

Figure 3-5. Relationship between the $\delta^{13}C$ for dissolved inorganic carbon and the concentration of phosphate in the water column of the northeastern Pacific Ocean. Data from Harmon Craig.

stand-in. The reason is that the CO_2 produced by respiration of the organic matter that falls to the deep sea has a 2 percent lower ^{13}C to ^{12}C ratio than does the resident dissolved inorganic carbon (DIC). As a result, respiration generates an inverse relationship between $\delta^{13}C^2$ and PO_4 content (see figure 3-5). For each micromole per liter of PO_4 released by respiration, the $\delta^{13}C$ in DIC is reduced by close to 1 per mil. Although complications exist when it comes to applying this relationship to the record kept in foraminifera shells, paleoceanographers have been able

$$^2\ \delta^{13}C = \frac{^{13}C/^{12}C)_{sample} - {}^{13}C/^{12}C)_{std}}{^{13}C/^{12}C)_{std}} \times 1000$$

to work their way around these shortcomings and have successfully reconstructed a picture of the phosphate distribution in the world ocean during the Last Glacial Maximum (LGM). Other than one difference that stands out like a sore thumb, the glacial phosphorus distribution was similar to that of the present. The difference involves the distribution in the deep Atlantic. In contrast to today's situation, in which the North Atlantic Deep Water mass with its nearly uniform phosphate content dominates the deep-water column, during glacial time a distinct boundary existed at a depth close to 2,500 meters, separating two distinct water masses. Above this boundary the deep water had a lower phosphate content than today's, and below 2,500 meters it had a higher phosphate content. The conclusion drawn from this is that deep water formed in the northern Atlantic during glacial time was not dense enough to sink to the bottom. Rather it spread out on top of water of Southern Ocean origin. In a sense, during glacial time a water mass akin to today's Antarctic Bottom Water had far greater success in penetrating the deep Atlantic. Instead of occupying only about 10 percent of the deep-water column, it occupied more like 50 percent.

The ^{14}C to C^{12} ratios in foraminifera shells carry important information concerning the rate of deep-sea ventilation during glacial time. This information is recovered from the radiocarbon-age differences between coexisting benthic and planktic foraminifera shells. These age differences can be compared to those computed from the ^{14}C to C^{12} ratios in prenuclear seawater.[3]

In today's ocean, these age differences have a pattern that reflects the residence time of water in the deep sea. In the tropical Atlantic, the differences are on the small side averaging close to

[3] As the result of hydrogen bomb tests, atmospheric CO_2 was tagged with excess radiocarbon. Much of this radiocarbon subsequently entered the upper ocean, raising its ^{14}C to C^{12} ratio well above the natural value.

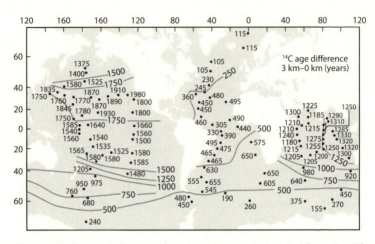

Figure 3-6. Apparent deep-water radiocarbon age calculated from the difference between the ^{14}C to C ratio for pre-1950 surface water and that for water at three kilometers depth. These ages reflect both the depletion of ^{14}C in polar surface waters and the loss by radiodecay in the deep sea. Even so, they make clear the large difference in residence time between deep water in the Atlantic and Pacific oceans.

350 years and in the tropical Pacific, the differences are much larger averaging about 1550 years (see figure 3-6). Neither of the ages, however, provides an accurate estimate of the residence time. Both are on the high side because part of each age difference reflects the deficiency in the polar-source water radiocarbon relative to tropical surface water. The actual residence time of deep water in the tropical Atlantic is more like 150 years and that in the tropical Pacific more like 1000 years. These differences must be kept in mind when considering the results based on measurements on coexisting benthic and planktic foraminifera shells.

The conditions influencing the production of deep water were seemingly quite different during glacial time. Sea ice was much more extensive in both polar regions. Sea level stood as much as 120 meters lower. The large northern ice sheets shifted the path of the atmosphere's jet stream. Latitudinal temperature gradients were stronger. Yet despite these changes within the measurement uncertainty (~200 years), the benthic-planktic age differences in tropical Pacific show no change over the last 25,000 years (i.e., from the last glacial maximum to the present).

This lack of significant change is perhaps less surprising when one considers the process which drives thermohaline circulation. It turns out that it is not what goes on in the polar source regions that matters. New deep water does not push its way into the deep sea. Rather, it is pulled into the deep sea. Heat stirred downward from the warm upper ocean into the interior gradually de-densifies the deep water. This allows more dense water at the polar surface to sink and underride the deep-water column. Hence it is the forces that drive the downward mixing of heat that interest us.

The ocean has a layer-cake structure, with each layer overlaid by a less dense one. Hence it takes energy to stir heat down into the interior. This energy is supplied by wind and by tides. Both cause the water to move. This movement creates friction between the layers as well as against bottom topography. The friction translates the energy associated with the motion into the energy needed to stir less dense water into more dense water. This being the case, the key factor is not how the polar regions changed but rather how the energy supplied by wind and tides changed. The constancy of benthic-planktic radiocarbon-age difference in the deep Pacific seems to be telling us that it did not change very much!

As might be expected from the glacial ^{13}C reconstruction, one would certainly expect to see differences between the glacial ^{14}C distribution in the Atlantic and today's. And indeed this is the case. The benthic-planktic age difference for the waters above 2,500 meters was even smaller than today's and that for the water below 2,500 meters was much greater—ranging all the way up to the sixteen-hundred-year difference seen in the deep Pacific. The smaller age difference for the Atlantic's upper reservoir suggests that a deep-water production rate similar to today's ventilated a reservoir roughly half the size of today's.

But, in a way, this information misses the point. The period of peak glaciation we have been discussing was not interrupted by Dansgard-Oeschger events. As had been the case during the Holocene, the Glacial Maximum appears to have been a time when ocean circulation was relatively stable. Therefore, once we get into the nitty-gritty of the millennial-duration episodes this subject will have to be revisited. What happened to the ocean circulation as a result of the abrupt changes that mark the onset and ending of the Younger Dryas and of each of the Dansgaard-Oeschger events?

CHAPTER 4

Puzzles

A serious challenge to the conveyor concept arose when it became clear that the impacts of both the Younger Dryas and the Dansgaard-Oeschger events were felt throughout the Northern Hemisphere and well into the tropics. If, as I had proposed, the impacts were the result of a shutdown of the heat carried to the northern Atlantic by the conveyor's upper limb, then only the region surrounding the northern Atlantic should have been affected and, in particular, the region downwind of the northern Atlantic (i.e., northern Europe). Early on, the failure to find a Younger Dryas signature in the pollen records from New England's bogs (or, for that matter, elsewhere in the world) appeared to be consistent with this line of thinking.

But then in the late 1980s and early 1990s evidence to the contrary started to pour in big-time. The first hint was turned up by my colleague Dorothy Peteet. She punched a core tube through several meters of the organic-rich mud in a bog formed in a depression in the glaciated surface of a diabase sill a short distance from the entrance to our Lamont-Doherty campus. The pollen record in this core revealed indisputable evidence for a Younger Dryas cooling. Although upwind and a bit too far south, with a stretch this could be passed off as a consequence of the shutdown of heat delivery by the conveyor. Dorothy's quest

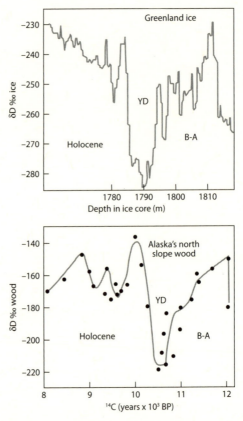

Figure 4-1. Comparison between the deuterium-to-hydrogen ratio (δD) record in cellulose from individual radiocarbon-dated twigs from Alaska's North Slope, as obtained by Caltech's Sam Epstein, and that for Dye 3 Greenland ice. Both records show a 60 per mil decline during the Younger-Dryas time interval.

for Younger Dryas impacts did not end there, however. She once again hit pay dirt in Alaska, where a sediment sequence revealed a climate reversal at the time of the Younger Dryas. Clearly, it was more than a stretch to attribute this change to a shutdown of northern Atlantic heat release. Furthermore,

Caltech's Sam Epstein produced a deuterium isotope record for cellulose from radiocarbon-dated twigs from Alaska's North Slope (see figure 4-1). It produced a record from Bølling Allerød through Younger Dryas into the early Holocene that beautifully replicated that in Greenland ice!

Before I had time for the significance of Dorothy's pollen records and Sam's deuterium record to fully sink in, a lightning bolt hit. Jim Kennett and a graduate student, Rick Behl, published a record from a high deposition-rate core from the Santa Barbara Basin recovered as part of the Deep Sea Drilling program.[1] I remember having breakfast on the patio of Jim's home, which overlooked this water body. He waved his arms and pointed, saying, "Soon I will have a long core from out there." I remember saying to myself, "So what?" It never occurred to me that the record from this faraway place would beautifully replicate that from the Greenland ice. So, when Kennett published his Santa Barbara Basin record a few years later, I was stunned. As shown in figure 4-2, the Younger Dryas and all the Dansgaard-Oeschger events were there!

In order to understand Kennett's record, a bit of background regarding the Santa Barbara Basin is necessary. It is a depression in the thirty-mile-wide channel that separates Santa Rosa Island from mainland California. The bottom water ventilating this depression spills in over a sill at a depth of about five hundred meters. The combination of low dissolved oxygen in eastern Pacific thermocline water and slow ventilation of the isolated depression causes its bottom-water oxygen content to hover close to zero. Consequently today's sediments are anoxic, and in the absence of O_2 benthic worms cannot survive. Hence

---

[1] Up until this time, most marine records were based on sediments that accumulated at only a few centimeters per thousand years. As worms stirred these sediments to depths of 6 to 10 centimeters, the imprint of millennial-duration events was lost.

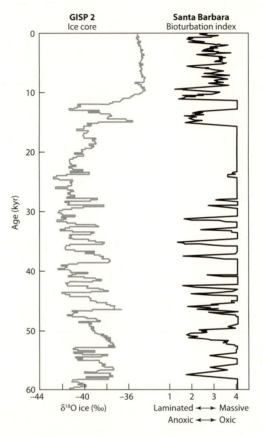

Figure 4-2. Comparison between the Greenland ice core ^{18}O to ^{16}O ratio record and the Santa Barbara Basin sediment core bioturbation record, as obtained in the laboratory of Jim Kennett of the University of California, Santa Barbara. Each of the ice core Dansgaard-Oeschger interstadial peaks is matched by an interval of laminated sediment (i.e., sediment anoxia) and each of the stadial peaks by a zone where the sediment was homogenized by worms. The timescale in the ice core is based on annual layer counting and that in the sediment core on radiocarbon dating.

no stirring occurs and the sediments' annual laminations are preserved.

What stands out clearly in this sediment is an alternation between layers that are annually laminated and layers in which these laminations have been disturbed by burrowing organisms. The message is clear: during the time intervals when burrowing occurred, the bottom water must have been reasonably well oxygenated, allowing worms to survive. Based on radiocarbon dating and also on the similarity to the Greenland ^{18}O record, Kennett demonstrated that the intervals of oxygenation corresponded uncannily to Greenland's cold episodes (i.e., the Younger Dryas, the LGM, and the Dansgaard-Oeschger stadial episodes) and the laminated intervals of anoxia to Greenland's warm episodes (the Holocene, the Bølling Allerød, and the Dansgaard-Oeschger interstadial episodes).

Although the circulation changes necessary to create better oxygenation are not clear, one possibility is that intermediate water was produced in the northern Pacific during the cold intervals. This would have required that the low salinity cap, which prevents this from happening today, was less of a deterrent. In the absence of ventilation in the northern Pacific, the thermocline water must come from outcrops in the Southern Hemisphere. During the long trip to the north, most of the O$_2$ is consumed.

A second ocean sediment record duplicating the Greenland pattern of events was published by Harmut Schulz and his German colleagues. It was based on organic-carbon-content measurements on a core off Pakistan in the northern Indian Ocean. During times of strong monsoon winds, upwelling fuels productivity, leading to greater rain of organic matter to the sea floor and hence greater O$_2$ depletion. Consequently, more organic matter accumulates. This was the first indication that the strength of the monsoons followed Greenland's tune. More about monsoons is to come.

One might ask why the records from Alaskan bogs and from Santa Barbara Basin and northern Indian Ocean sediments came as such a shock when we already knew from the Greenland methane and dust records that the impacts were widespread. The answer is that the methane record came only later. Whereas the dust record was available from the beginning, it remained something of a curiosity. It was not until Lamont's Pierre Biscaye showed that the dust originated in the faraway Asian deserts that its significance sank in.

While these far-field records posed a threat to the conveyor idea, the record from sediments of the Cariaco Basin in the Caribbean Sea off the coast of Venezuela gave it a boost. As is the Santa Barbara Basin, the Cariaco is a closed depression in a channel separating the South American mainland and offshore islands. It is ventilated by water spilling over a shallow sill. Through most of its history, its sediments remained anoxic, allowing annual layering to be preserved. Because of changes in the input of detritus from the adjacent land however, the packets of annually laminated sediments representing intervals of discreet climate differ in color. In particular, the time interval represented by the Younger Dryas is lighter in color than that for the Holocene-age sediment that lies above it and the Bølling Allerød–age sediment that lies below it. Furthermore, the boundaries of these color transitions are very sharp, just as they are in the ice core record (see figure 4-3)!

More important than the fact that it represents yet another spot on the globe that responded to the Younger Dryas and Dansgarard-Oeschger events is that the Cariaco record provides evidence that the ocean's radiocarbon distribution was affected (see figure 4-3). Konrad Hughen, as part of his PhD

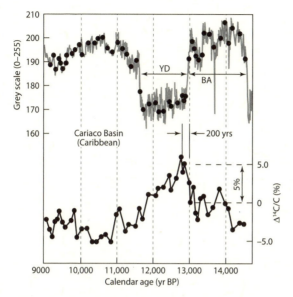

Figure 4-3. Based on counts of annual layers, Konrad Hughen was able to construct an absolute chronology for a sediment core from the Cariaco Basin off Venezuela. Using this chronology, he converted radiocarbon measurements on planktic foraminifera to ¹⁴C to C ratios at the time the shells formed. As the sediment deposited during the Younger Dryas is lighter in color than that deposited during the preceding Bølling Allerød or the subsequent Holocene, the boundaries of this cold episode are well defined. As can be seen, during the first 200 years of the 1,350-year duration YD, the ¹⁴C to C ratio in Cariaco Basin surface waters (and presumably also in the atmosphere) rose by 5 percent. This rise is thought to be the result of a partial shutdown in the delivery of ¹⁴C to the deep sea. As a result, the newly produced ¹⁴C atoms were backlogged in the atmosphere and upper ocean. After 200 years had passed, somewhere in the polar regions renewed deepwater production must have commenced, which caused the backlogged ¹⁴C to be drained back down. The circles represent the depths at which ¹⁴C measurements were made.

thesis at the University of Colorado, was able to take advantage of the absolute timescale provided by the annual layers present in Cariaco sediments. He showed that during the first 200 of the 1,350-year-duration Younger Dryas the radiocarbon content of the atmosphere rose by about 5 percent, and that during the remaining 1,150 years it coasted back down again. To avoid the necessity of counting the varves through the entire 11,700 years of post–Younger Dryas time, Hughen was instead able to match the wiggles in a very detailed set of radiocarbon measurements on planktic foraminifera shells he picked from the Cariaco sediment with an equally detailed set of radiocarbon measurements available for the dendrochronologically dated European tree-ring series. This being accomplished, he started his layer counts in the early Holocene and extended them back through the Younger Dryas into the Bølling Allerød. In this way, he had to count only a few thousand varves rather than more than thirteen thousand! He then used radiocarbon ages on shells from his counted section to document disparities between radiocarbon years and varve years. In this way, he was able to extend the record of the magnitude of the offset between radiocarbon ages and calendar ages back in time well beyond that based on European trees. At that time, the tree-based offset record extended only to the base of the Holocene. While trunks of Holocene age were abundant in sand and gravel deposits in European rivers, those of Younger Dryas age were as rare as hens' teeth.

To me, Hughen's results were very exciting. The 5 percent rise in the ^{14}C to C ratio in Cariaco surface waters that occurred during the first two hundred years of Younger Dryas time was consistent with my idea that a conveyor shutdown caused the Younger Dryas. In today's ocean, despite the fact that it supplies only half of the new deep water, ventilation from the northern Atlantic supplies about 80 percent of the radiocarbon atoms reaching the deep sea. During its northward trip, waters of the conveyor's upper limb are able to pick up radiocarbon from the atmosphere but the deep water descending along the margins of the Antarctic continent contains very little new radiocarbon. The reason is that the water upwelling under the ice has very little chance to exchange carbon atoms with the atmosphere. No sooner has it reached the surface than it is cooled and sent back down. Cooling occurs in months. CO_2 exchange requires the better part of a decade.

So, as a result of the shutdown of the northern Atlantic supply route, the newly produced radiocarbon atoms entering the ocean from the atmosphere were backlogged in the upper ocean. As a consequence, the ^{14}C to C ratio in upper ocean and atmosphere carbon slowly rose and, of course, that in the deep sea correspondingly declined (i.e., the same number of ^{14}C atoms were present but they were distributed differently).

A simple example helps to understand this. Consider what would happen if the lower two-thirds of the ocean were suddenly cut off from communication with the upper one-third. The ^{14}C to C ratio in the deep sea would begin to decrease at the decay rate of radiocarbon (i.e., ~1 percent per 80 years). As a number of ^{14}C atoms matching this loss would be added to the upper ocean, its ^{14}C to C ratio would rise by 2 percent every 80 years. After 200 years, the ^{14}C to C ratio in the deep sea would have dropped by 2.5 percent and that in the upper ocean (and atmosphere) would have risen by 5 percent.

But why, after rising during the first 200 years of Younger Dryas time, did the ^{14}C to C^{12} ratio then coast back down again? As you remember from chapter 3, the energy added to the ocean by the winds and the tides is dissipated by mixing heat down into the ocean's interior. As this happens, the bottom water's density is reduced, allowing new deep water to push its way underneath. The point is that the shutdown of deepwater formation would not disrupt vertical mixing. Rather, this process would continue as before. Hence, unless the production of new deep water somewhere else in the world replaced the deficit created by the shutdown, the density reduction would increase. Presumably after 200 years something changed, allowing more deep water to form and, in the process, to a gradual draw down of the ^{14}C backlogged in the upper ocean as the result of the shutdown. Although this explanation satisfied me at the time, it turned out to have problems, as have many aspects of this subject. Hence, the story does not end here.

While the Cariaco ^{14}C record gave the conveyor idea a boost, Richard Seager, a colleague of mine here at Lamont, made a discovery that in his mind relegated it to the waste bin. He even went so far as to show a tombstone in his lectures inscribed, "Here lies Wally's conveyor belt . . ." His attack was based on model simulations carried out using a finer grid than those which previously lent support to the idea that northern Europe experienced moderate winters as the result of heat given off by the conveyor. These more advanced models showed that this was not the case. Rather, the warmer temperatures were related to a bend in the atmosphere's jet stream generated by the underlying mountain ranges. Although I accepted Seager's model results, a new wrinkle to be discussed in the next chapter bailed out my idea. Otherwise there would be no reason to write this book.

Before we leave this chapter, we must examine the record in the Southern Hemisphere. Did it experience Younger Dryas and Dansgaard-Oeschger events? The obvious place to start the search would be in Antarctic ice. The early record from the Byrd station core failed to show these events. Although these events are also absent in subsequent cores, attention focused on a plateau in the warming that interrupted the demise of the last glacial period. Instead of rising smoothly, about halfway through the transition the ^{18}O to ^{16}O ratio stopped rising and

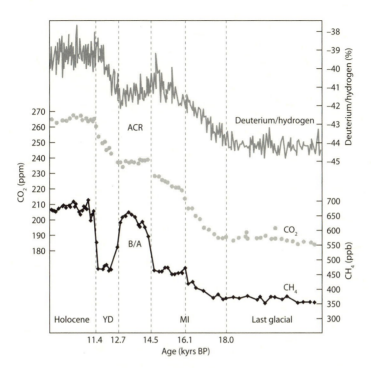

Figure 4-4. An Antarctic record of the deuterium-to-hydrogen ratio (δD) in the ice and of the CO_2 and CH_4 concentrations in the air trapped in bubbles. Based on the methane concentration, the depth intervals corresponding to the Bølling Allerød (BA) and to the Younger Dryas (YD) are defined. As can be seen, during the time of the Bølling Allerød, the warming of Antarctica and the rise in the atmosphere's CO_2 content stalled. This interval is referred to as the Antarctic Cold Reversal (ACR). Then during the time of the Younger Dryas, the Antarctica's warming and the atmosphere's CO_2 rise resumed. The ages for the transitions points have been adjusted to conform to the [230]Th chronology for the Huly Cave [18]O record as obtained by the University of Minnesota group.

remained nearly constant for about two thousand years (see figure 4-4). Toward the end of this plateau, it underwent a small cooling that gave rise to its designation as the "Antarctic Cold Reversal." Until the methane was measured in these cores, this plateau interval was tentatively linked to the Younger Dryas. I say tentatively because unlike Greenland ice, that in Antarctica lacks annual layering. Hence the dating was approximate. This chronology problem was firmed up only when methane records were obtained in both Antarctic and Greenland ice cores. This allowed the Greenland layer-counting chronology to be applied to the Antarctic ice-core record. The big surprise was that the methane drop that corresponded to the Younger Dryas cooling in Greenland ice clearly postdated the plateau in the Antarctic stable isotope records (see figure 4-4). Hence this plateau was shown to correspond in time to the Bølling Allerød warm interval and not to the Younger Dryas.

At a meeting held in our Lamont Hall, Jerome Chappallaz from the Grenoble ice core laboratory presented his brand new methane results. I was dumbfounded but recovered quickly when I realized that Jerome had discovered evidence for what I termed the bipolar seesaw. Comparison of the ^{18}O records at opposite ends of the planet for the period of deglaciation revealed a reciprocal relationship. About 18,000 years ago, Antarctica started to warm. This warming continued until about 14,500 years ago, when it stalled. By contrast, Greenland remained quite cold during this interval and 14,500 years ago underwent a large sudden warming, initiating the Bølling Allerød, which lasted until 12,700 years ago. During this 1,800-year period, the warming in Antarctica plateaued. Then 12,700 years ago, when the Antarctic warming resumed, Greenland's temperature plunged to its Younger Dryas low value. Antarctic's warming was completed during the 1,300-year Younger Dryas time interval.

As discussed in chapter 3, deep-water formation is driven by the dedensification of the abyssal ocean by the downward penetration of heat. Therefore, if for some reason production is cut off in one region of the polar ocean, then on a timescale of no more than a century or two it will be initiated elsewhere. The seesaw involves an interplay between deep-water formation in the northern Atlantic and that in the Southern Ocean. Furthermore, as the heat released by this process is important to the atmosphere's budget, the ocean's seesawing leads to an anti-phasing between climate events in the Northern and Southern Hemispheres.

At this point, it is worth mentioning yet one more false trail I followed. On New Zealand's South Island there is a well-preserved late glacial moraine[2] (called the Waiho Loop) produced by the Franz Joseph Glacier (see figure 4-5). It appeared to be a perfect candidate for a Younger Dryas re-advance. Indeed, in 1994 George Denton and Chris Hendy published an article in *Science* in which, based on twenty-four radiocarbon-dated wood samples lodged on a rock knob just inside the moraine, they made a case that indeed it was a Younger Dryas event. Because of this finding, I became convinced that the boundary between the two climate worlds lay to the south of New Zealand. This meant that only the Southern Ocean followed the Antarctic deglacial pacing and the rest of the planet followed Greenland's.

A nagging doubt remained, however. The majority of the radiocarbon ages were a few hundred years too old. In other words, the wood appeared to have formed before the onset of the Younger Dryas event. Denton and Hendy passed this off by admitting that indeed most of the wood grew before the Younger Dryas, but was subsequently transported down valley by the advancing Younger Dryas glaciers.

Years later, this explanation was challenged. Critics claimed that the trees grew on their host rock knob and were killed only when the glacier reached close to its outer limit. Denton

---

[2] Moraines are piles of debris either plowed into place by a glacial advance or left behind as the ice melted during a pause in its retreat.

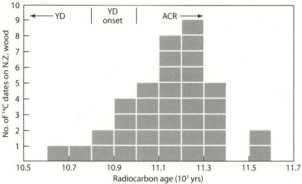

Figure 4-5. The moraine shown in this photograph was produced by an advance of the Franz Josef Glacier down the western flank of the New Zealand Alps. Initial dating of wood samples from the Canavan Knob (located just behind the moraine itself) yielded radiocarbon ages that span from the late Bølling Allerød to the Younger Dryas, raising the question of whether this glacial advance was a Younger Dryas correlative or an event correlating with the Antarctic Cold Reversal. Recently, George Denton obtained radiocarbon ages on newly collected samples of Canavan Knob wood. These ages (obtained on bark) suggest that the culmination of this advance occurred during the Antarctic Cold Reversal. The importance of the bark is that its preservation eliminates the possibility that the wood was "pre-aged" before transport to the knob. Data and photograph courtesy of George Denton of the University of Maine.

responded by recollecting wood samples with bark intact and had the bark rather than the wood radiocarbon dated. His logic was that the bark would not have survived transport down the steep valley. So his new samples must have grown in place. Furthermore, as the bark was the last growth, its choice eliminated the possibility that the wood was from the inside of a tree. When the results came back, all but one piece of bark predated the Younger Dryas. The range of results was identical to that for the earlier set. So the interpretation had to change. The glacier advanced during the Antarctic Cold Reversal rather than during the Younger Dryas. Hence I no longer believe that the boundary between the two climate regimes was located to the south of New Zealand.

Before continuing the discussion on the boundary's location, one more piece of evidence must be considered because it provides the key that allows us to answer this chapter's puzzle: namely, how did a conveyor shutdown lead to such widespread Younger Dryas and Dansgaard-Oeschger impacts? This evidence will be the subject of the next chapter.

CHAPTER 5

Hot Clues

I n 1988, Hartmut Heinrich, a young German marine geologist, published an article that turned out to hold the key to
solving the puzzle posed in the last chapter. He described six
discreet layers of debris in the last glacial section of an eastern
North Atlantic sediment core (see figure 5-1). The composition
of the coarse fraction grains in these layers differed dramati-
cally from that of the ambient sediment. They were dominated
by mineral grains rather than by foraminifera shells. Heinrich
noted that the bases of four of these layers were razor sharp,
suggesting an abrupt onset for their emplacement. An acquain-
tance offered to carry out potassium-argon measurements in
an attempt to ascertain their provenance. The results showed
that the lithic material in the layers had ages close to two bil-
lion years, while for the ambient sediment the age was only several hundred million years. Based on these observations, young
Heinrich drew the bold conclusion that the layers were created
by the melting of iceberg armadas launched from Canada's Laurentide ice sheet!

Heinrich's article went largely unnoticed, perhaps because it
didn't fit nicely into the prevailing view that growth and retreat

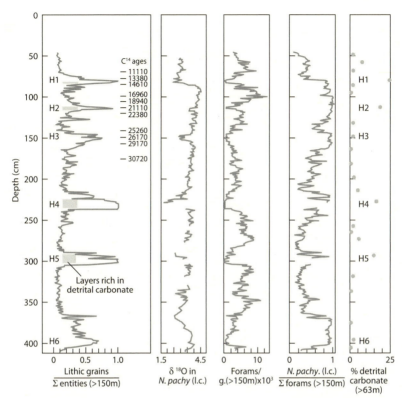

Figure 5-1. Heinrich-event records covering the last glacial period as recorded in Ocean Drilling Program core 609 (50°N, 24°W). The left-hand
panel shows the ratio of lithic fragments to foraminifera shells plus lithic
grains (i.e., total entities) in the >150 μm size fraction. Shown by black
bars are the layers bearing detrital limestone. Also shown are the radiocarbon ages. In the other four panels are shown, respectively, the $\delta^{18}O$ record for *N. pachyderma* (l.c.), the number of foraminifera shells per gram
of sediment, the ratio of *N. pachyderma* (l.c.) to total foraminifera shells,
and the percent detrital limestone fragments in the lithic fraction. The ab-
breviation *l.c.* refers to left-coiling *N. pachyderma*, a species that today is
found only in waters with temperatures lower than 5°C. These results were
obtained by Gerard Bond and his Lamont-Doherty group.

of ice sheets was smoothly orchestrated by Milankovitch's orbital cycles. I was one of a small minority who suspected that Heinrich was on to something really important, but for the life of me I couldn't figure out what it was. His six events, separated from one another by seven or so thousand years, occurred during the same time interval as the Dansgaard-Oeschger events. When I examined the Greenland ice-core record, however, I could find no features corresponding to Heinrich's armadas.

And so it stood until one day Lamont's Gerard Bond asked for my opinion of a proposal he was preparing. To make his point, he included a reproduction of the photograph of sections of a Deep Sea Drilling core raised in the northeastern Atlantic. Although he didn't mention the distinct thin white layers that stood out from the buff-colored ambient sediment, they were all that I saw. I went running over to Bond's office and asked if, by chance, these white layers could be Heinrich's ice-rafted debris layers. Because Bond's interest lay in sedimentary sequences of Paleozoic age exposed in the mountains of British Columbia, he had never seen Heinrich's article. But he humored me, and we went to Lamont's sediment archive and took a sample of the white material. Lo and behold, it was loaded with ice-rafted debris and foraminifera shells were few and far between. Under Bond's supervision, Elizabeth Clark and Millie Klas, who worked in my group, carefully sampled the entire glacial section of the core and made determinations of the ratio of lithic grains to foraminifera shells. Soon they had beautifully duplicated Heinrich's record (see figure 5-1). Captivated by this finding, Gerard Bond put aside his research on Paleozoic sediments and began what turned out to be more than a decade of effort studying ice-rafted debris in northern Atlantic sediments.

We were joined in these efforts by Sidney Hemming, a geologist interested in using the radiogenic isotopes of lead,

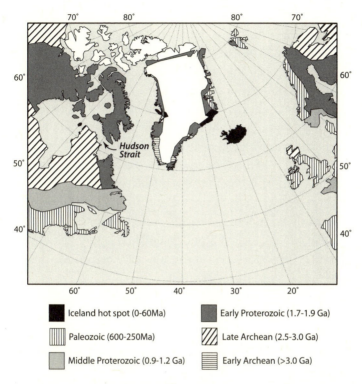

Figure 5-2. Map, created by Sidney Hemming of Lamont-Doherty, showing the geologic age provinces that underlay the ice sheets of the last glacial period. Consistent with a Hudson Strait origin, fragments of Early Proterozoic rock dominate Heinrich layers 1, 2, 4, and 5.

strontium, and neodymium to constrain the provenance of the mineral grains in sediments.[1] She was able to show that the debris in four of the six Heinrich layers was dominated by debris eroded from Canada's Archean-age Churchill Province (see figure 5-2), which surrounds the northern portion of the Hudson

---

[1] *Provenance* refers to the source of the material found in a sediment.

Bay. This finding led the University of Chicago's Doug McAyle to propose a "binge and purge" hypothesis for the origin of these armadas. Trapped by the insulation provided by the overlying ice, heat diffusing up from the Earth's hot interior gradually warmed the base of the ice (the binge). When it reached the melting point, lubrication by water allowed the ice to collapse and push its way through the Hudson Straits into the Atlantic Ocean (the purge), creating an armada of bergs.

Examination of many sediment cores allowed mapping of the track followed by each of these armadas. Indeed, as expected if the bergs were launched from the mouth of the Hudson Straits, each debris layer thinned to the east (see figure 5-3). Cores proximal to the Hudson Straits had debris layers many tens of centimeters thick, whereas those closer to the British Isles had layers only a few centimeters thick.

As a demonstration that the layers accumulated rapidly, Jerry McManus (at Lamont) made measurements of the amount of excess ^{230}Th they contained. The idea was that, as the magnitude of the rain of this particle reactive isotope was fixed by the amount of its parent uranium dissolved in the overlying water column, the concentration in the sediment should be inversely proportional to the accumulation rate. The results were consistent with expectation; the concentration of ^{230}Th in the Heinrich debris was far smaller than that in the ambient sediment.

So, it turned out that Harmut Heinrich had it right. The light-colored layers with few foraminifera and much ice-rafted mineral grains dropped from melting icebergs. Furthermore, each armada was launched from the Hudson Straits presumably as the result of a collapse of the Hudson Bay lobe of the Laurentide ice sheet. These layers now bear the name of their discoverer. They are referred to as Heinrich layers.

Attempts to constrain the volume of ice comprising each armada have been inconclusive. A pronounced dip in the ^{18}O to

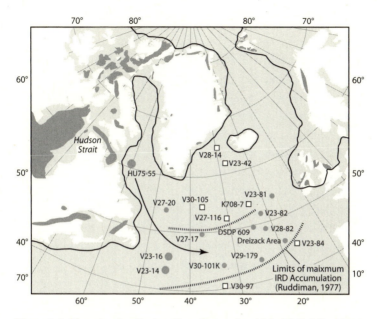

Figure 5-3. Map showing the location of cores in which detrital limestone-bearing Heinrich layers have been identified (solid circles) and of cores in which they are absent (open squares), as created by Sidney Hemming of Lamont-Doherty. The size of the circles is proportional to the thickness of the layers. The black patches on the land masses denote occurrences of sedimentary limestones that might serve as sources for the detrital calcite. The solid line shows the maximum size of the glacial ice sheets. The arrow shows the likely path taken by the Heinrich ice armadas.

^{16}O ratio in foraminifera shells from within and just above the layers demonstrates that heavy isotope–deficient glacial melt-water was present (see figure 5-1): estimates of the amount of ice required to produce these dips is strongly dependent on the duration of the melting episode and the rate of lateral mixing of the meltwater. As both of these are poorly constrained, estimates of the volume of ice (expressed as the resulting rise in sea level) range from as little as twenty centimeters to as much as twenty meters.

One might ask why the total amount of mineral debris dropped to the sea floor by each armada doesn't tell us something in this regard. The problem is that we know from Greenland ice cores that only the ice very close to the bedrock contains the coarse debris similar to that in Heinrich layers. This being the case, the very large amounts of sea floor debris has led to creative thinking regarding how to incorporate the required large debris loads into basal ice. In any case, the amounts of debris do not constrain the size of the armadas.

One approach to estimating their size has yet to be adequately exploited. Ice caps contain both the cosmic ray–produced ^{10}Be- and ^{3}He-rich interstellar dust grains. The rain rate of both of these entities onto ice sheets has been well documented. As both ^{10}Be and interstellar grains released as the bergs melted should have been deposited locally on the sea floor, inventorying these entities in the debris layers holds promise as a way to narrow the range in estimates of the amount of fresh water released as a result of the melting of each Heinrich armada. Such measurements are currently being conducted.

As the sudden addition of large amounts of fresh water to the surface of the northern Atlantic is the perfect way to squelch deep-water formation in the northern Atlantic, one would expect that impacts of the Heinrich armadas should show up in Greenland ice and, for that matter, in Santa Barbara Basin sediments. But they don't. One possible explanation is that the armadas triggered six of the twenty or so Dansgaard-Oeschger events. But if so, their consequences were no different from those of whatever triggered the remainder of the events.

It was not long, however, before two records appeared that showed strong responses at the times of the Heinrich events but no distinguishable responses at the times of Dansgaard-Oeschger events. Both records are from the tropics.

One is a record from a lake in central Florida. Pollen grains isolated from its sediments show an alternation between layers rich in oak and those rich in pine. Eric Grimm of the Illinois Geological Survey obtained a series of radiocarbon dates on organic matter for these sediments and was amazed to find that the age of each pine layer fell close to that of a Heinrich event (see figure 5-4). Puzzling, though, was the observation that the length of the pine intervals was close to that for the oak intervals. As there was no indication of any change in sediment accumulation rate between these intervals, the suggestion is that the duration of the periods when pine trees dominated was similar to that of periods when oak trees dominated. Hence, although the first record to reveal climatic impacts correlated with Heinrich events, the several-thousand-year duration of the periods of pine domination was difficult to understand, especially since each of these time intervals was punctuated by at least one Dansgaard-Oeschger event.

The second record showing only Heinrich impacts was from a sediment core taken from the continental margin off eastern Brazil. Helge Arz, a young German paleoceanographer, conducted measurements of calcium and iron on this core. What he found was that while the ambient sediment was rich in calcium, its deposition was interrupted by six layers where the calcium content dropped and iron became abundant (see figure 5-4). His interpretation was that the calcium-rich layers were normal marine ooze rich in $CaCO_3$ produced by foraminifera and coccolithophoriads and the iron-rich layers represented brief intervals of high river runoff leading to the rapid deposition of continent-derived soil detritus. As eastern Brazil is currently a dry area lying beyond the reach of Amazonian rains, these intervals rich in continental detritus must represent times of greatly increased precipitation! Arz obtained radiocarbon dates that documented that these rainy intervals came close to the times of Heinrich events. As in Florida, there was no hint that Dansgaaard-Oeschger events created similar episodes.

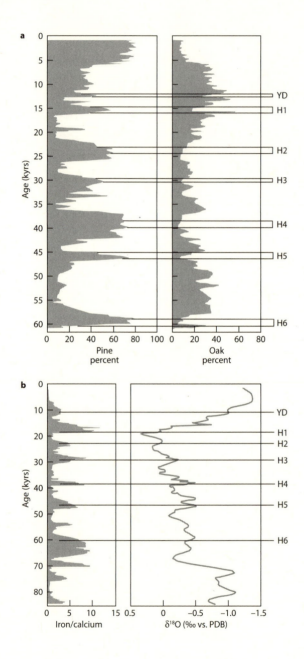

Not discouraged by the absence of any obvious Heinrich-related features in the Greenland ice-core records, Gerard Bond used radiocarbon dates on North Atlantic sediment (corrected for the calendar offset[2]) to place the Heinrich events into Greenland's ^{18}O record. He found that there appeared to be a subcycle in the amplitude of the oxygen-isotope changes. At the time of each Heinrich event, this subcycle appeared to be reset (see figure 5-5). Although not totally convincing, at least there was a suggestion that the ice armadas did, after all, leave an imprint in Greenland ice.

With this in mind, the Schulz Indian Ocean record was reexamined. Clearly, at the times of the Heinrich events the drops in organic carbon content were larger than those associated with the Dansgaard-Oeschger events. The likely explanation for the fluctuations in organic carbon content seen in this record is that they reflect changes in the strength of the winds responsible for Indian monsoon rains. Again, more on this later in this book.

One other record is worth mentioning. It comes from a high accumulation-rate sediment core taken in the Atlantic Ocean

[2] Due to temporal changes in the rate of production, radiocarbon ages deviate from calendar ages. The offsets were as large as three thousand years during peak glacial time.

Figure 5-4

a. Pollen record obtained by Eric Grimm of the Illinois Geological Survey from Lake Tulane in central Florida. The time of each Heinrich event (and the Younger Dryas) is marked by a prominent increase in the abundance of pine pollen and a corresponding reduction in oak pollen.

b. Records of oxygen isotope for *G. sacculifer* and of iron to calcium ratio in bulk sediment, obtained by Helge Arz while at Germany's Bremen University, for a sediment core from 770 meters depth off Fortaleza, Brazil (4°S, 38°W). Times of high iron to calcium ratio (i.e., soil detritus to marine $CaCO_3$) correlate with Heinrich events (and the Younger Dryas).

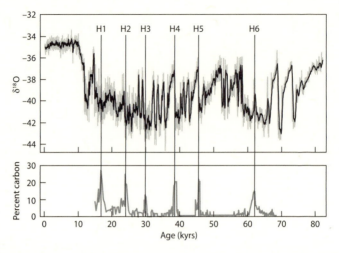

Figure 5-5. Gerard Bond's placement of the Heinrich ice armadas in the sequence of Dansgaard-Oeschger events as recorded in the GISP2 ice core. As can be seen, there appears to be a subcycle of weakening D-O events that culminates in a Heinrich outbreak.

offshore from Portugal. Edouard Bard, a French geochemist, showed that both Dansgaard-Oeschger and Heinrich events were recorded. Both appeared as marked coolings, but those at the times of the Heinrich events were more pronounced. Hence, as in the northern Indian Ocean core, the imprints of the ice armadas were stronger than those of Dansgaard-Oeschger events.

In summary, the records in Florida, in Brazil, and in the northern Indian Ocean all point to the same thing. Somehow, the Heinrich armadas triggered changes in tropical rainfall. As we shall see in the next chapter, this turned out to be the big clue that led to the explanation as to why the disruptions of the Atlantic's conveyor operation led to such widespread impacts.

CHAPTER 6

The Solution

The solution to the puzzle posed by the widespread impacts caused by disruptions of the Atlantic's conveyor operation came in two steps. Step one involved observations made by George Denton on Greenland's glacial moraines and step two involved model simulations carried out by John Chiang, an atmosphere scientist at the University of California, Berkeley.

In 2001, the late Gary Comer was able to navigate his yacht, *Turmoil*, through the Northwest Passage without any interference from ice. In so doing, he experienced first-hand the effects of a subject that would hold his attention for the rest of his life, namely, abrupt climate change. To learn more about this, he "adopted" a number of us who do paleoclimate research and offered to provide us with assistance. George Denton suggested that Comer use the *Turmoil* as a base for a field study of moraines in Greenland purported to be of Younger Dryas age. Comer agreed, and in the summer of 2003 Denton led a group to study moraines halfway up Greenland's east coast in the area around Scoresby Sund (see figure 6-1). The group mapped the moraines and obtained samples from glacial erratics for ^{10}Be

Figure 6-1. Map of Greenland showing the locations of the Greenland ice cores and also of Scoresby Sund, where Denton made his Younger Dryas snowline reconstruction. The Severinghaus measurements were made on ice from the GISP2 core.

dating.[1] By mapping the moraines, the magnitude of the depression of the snowline associated with glacial advance could be established. As the extent of the glaciers at high latitudes is set by summer conditions, Denton was able to show that the summer cooling (relative to today) required to create these moraines was a modest 4° to 6°C.

When the ^{10}Be dates confirmed earlier radiocarbon measurements that had tentatively placed the moraines as Younger

---

[1] Some of the neutrons produced by the cosmic rays bombarding our Earth make it to the surface and break up the nuclei of atoms in rocks. The radioactive ^{10}Be atoms created from the breakup of oxygen nuclei in quartz grains prove ideal for dating. The longer a rock is exposed at the Earth's surface, the more ^{10}Be is produced in its quartz grains.

Dryas in age, Denton was mighty puzzled. Although his summer cooling estimate agreed with those based on pollen and beetle remains in Younger Dryas–age deposits from Scandinavia and Great Britain, they were in strong conflict with temperature estimates based on measurements made on ice from the GISP2 ice core drilled on Greenland's polar plateau directly inland from Scoresby Sund. Measurements made by Jeff Severinghaus suggested that the mean annual temperature during the Younger Dryas was a whopping 16°C colder than now.

Before discussing the cause of the difference between Denton's ~5°C Younger Dryas cooling and the Serveringhaus' estimate of ~16°C, let us consider how the latter was obtained. The ^{18}O estimate for Greenland's peak glacial cooling was about 10°C and the difference between the Holocene and the Younger Dryas temperature was about 8°C, that is, less discordant with Denton's moraine-based estimate of 5°C. So, why was it that the ^{18}O-based glacial estimate was put aside in favor of the much larger change?

The answer is that it was "trumped" by a more direct estimate—that based on a down-hole temperature log. Once the drilling to bedrock was complete, the hole left open by the removal of the ice was filled with kerosene. As the density of this fluid is nearly the same as that of ice, its presence prevented the hole from closing as the result of flow of the ductile ice. Once the temperature of the kerosene had equilibrated with that of the surrounding ice, an electronic thermometer was lowered into the hole and run slowly up and down, creating a highly precise temperature profile from surface to bedrock.

The next step was to use the shape of this profile to reconstruct the air temperature above the ice during glacial time. This required correcting for the down core warming generated by the diffusion during Holocene time of heat down through the ice. This so-called deconvolution is a complex task that must take into account accumulation rate of the ice, the thinning of the ice resulting from lateral flow, and the time history of the *shape* of temperature change (based on the ^{18}O record). It allows the surface temperature record to be reconstructed. The key feature in the measured profile that permits this is the temperature minimum well down in the hole (see figure 6-2). Although the downward diffusion of heat has gradually eroded this feature, enough remains to permit the average temperature during glacial time to be reconstructed. I say "average" because the imprints of shorter-duration features such as the Dansgaard-Oeschger events and the Younger Dryas event have been smoothed beyond recognition. The important point is that to explain the magnitude of the cold bump, a glacial temperature of somewhere between 20° and 25°C colder than today's is required!

But how can the ^{18}O-based estimate be so wrong? The answer lies in the seasonality of snowfall. As summers are warmer than winters, the ^{18}O to ^{16}O ratio in summer snows is less negative than that for winter snow. Today the accumulations of summer and winter snow are similar, so the isotope thermometer provides a mean annual temperature. But what if, during the cold glacial time, the snowfall was much greater during summer months than during winter months? If so, the ^{18}O-based temperature estimate would be biased toward summer temperatures. Aha! As you can see, we are on the track to an explanation. But before jumping to any conclusion, we must consider one other piece of information.

Jeff Severinghaus came up with a very clever way to estimate the magnitude of the abrupt warmings that marked the termination of each Dansgaard-Oeschger event and also that of the Younger Dryas. His approach was to make very precise measurements of the ^{15}N to ^{14}N ratio in the N_2 gas trapped in

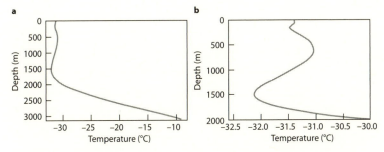

Figure 6-2. Temperature profile in the hole created by the recovery of the GISP2 ice core as measured by Kurt Cuffy of the University of California, Berkeley. The two minima shown in the blowup of the upper two kilometers of the record are remnants, respectively, of the peak glacial cooling (22,000 years ago) and the Little Ice Age cooling (~300 years ago). The base of the ice is warmed by geothermal heat diffusing up from beneath the ice cap.

ice-core bubbles. What interested Jeff was a process referred to by physicists as thermal diffusion. If a gas is subjected to a thermal gradient and not allowed to undergo convective mixing, a small separation between the heavy and light molecules takes place, enriching the heavy entities at the cold end and the light ones at the warm end. This process was explored during World War II as a way to enrich fissionable ^{235}U for use in atomic bombs.

Jeff viewed the roughly eighty-meter-thick firn as the ideal place for this process to occur, as convection of the gas through the connected pores does not occur; only molecular diffusion takes place. Furthermore, he reasoned that as a result of the abrupt warming that occured at the end of the Younger Dryas cold period, a thermal gradient would have been set up in Greenland's firn. The temperature at its surface would immediately rise but that at its base would initially remain unchanged. As a century or more would be required for the warming to penetrate to the base of the firn, Jeff reasoned that there should be a temporary enrichment of heavy nitrogen gas (i.e., $^{15}N\ ^{14}N$) in the bubbles closing off at the base of the firn. Furthermore, the magnitude of this enrichment would depend on the magnitude of the abrupt warming. So he carried out the painstaking measurements and was rewarded by finding the predicted temporary increase in ^{15}N to ^{14}N ratio (see figure 6-3). The

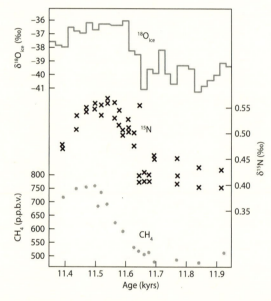

Figure 6-3. Plots across the end of the Younger Dryas of the ^{18}O to ^{16}O ratio in ice and of the ^{15}N to ^{14}N ratio in N_2 and the methane concentration in the air trapped in bubbles, as obtained by Jeff Severinghaus of Scripps Institution of Oceanography. The elevated N_2-isotope ratio in Younger Dryas air is the result of gravitational settling in the firn layer. The abrupt rise in this ratio at the end of the Younger Dryas is the result of thermal diffusion induced by the abrupt warming. As can be seen, the rise in atmospheric methane begins very close to the time of the abrupt warming. The offset between the ice record and the bubble record was removed by matching the sharp rises in ^{18}O and ^{15}N.

magnitude of the enrichment corresponded to an 8°C warming. As only about half of the post–Younger Dryas [18]O change occurred abruptly (the remainder taking place over hundreds of years), Jeff estimated the total post–Younger Dryas warming to be more like 16°C.

This being the case, George Denton was faced with explaining why his mountain glaciers recorded only a 4° to 6°C difference between the Younger Dryas and the Holocene. It didn't take him long to realize that the answer was that Younger Dryas winters were far, far colder than today's. For example, if they were 27°C colder than today's, then when averaged with the 5°C colder summer temperatures, the mean annual cooling would be 16°C: the value Jeff obtained!

But how could this be? Only in landlocked Siberia does such a large temperature swing between summer and winter occur. Surrounded by ocean water, this could not be the case for Greenland. Denton reasoned, what if there were ice instead of water? That was it; if, under the conditions that prevailed during glacial time, the delivery of heat by the conveyor were to be cut off by a flood of fresh water, then during the following winter the ocean surrounding Greenland would freeze over (see figure 6-4). Once this happened, no heat would escape to the overlying atmosphere. Furthermore, instead of absorbing sunlight, the ice-covered ocean would reflect it back to space. Winters in Greenland would become like those in Siberia. Of course, just as the snows that blanket Siberia melt away under the summer sun, the ice covering the sea around Greenland would have melted back, allowing more moderate summers. Furthermore, as is the case in today's Arctic, the fresh-water lid would have remained from year to year, preventing the renewal of deep convection and permitting sea ice to reform each winter.

So it all came together. The winters were so cold in Greenland that little snow fell. Hence, the [18]O thermometer recorded

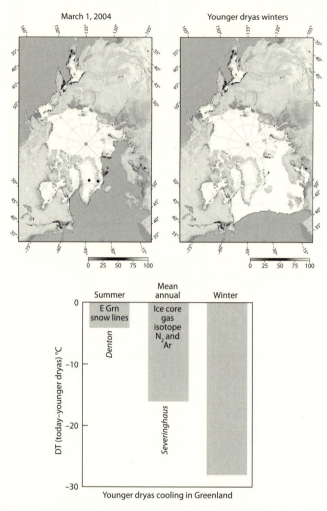

Figure 6-4. Maps comparing the postulated winter sea ice cover of the northern Atlantic region during the Younger Dryas with that for the year 2004 AD. Also shown are the reconstructions of the Younger Dryas summer cooling (Denton) and of the Younger Dryas mean annual cooling (Severinghaus). In order to explain the difference between these two estimates, it is necessary to call on a very large winter cooling. The dots designate the location of the ice core and of Scoresby Sund.

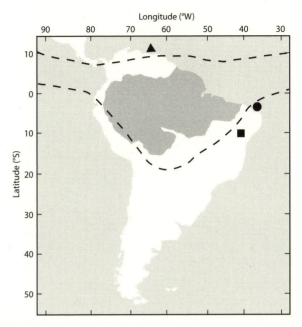

Figure 6-5. Today's northern and southern limits of the seasonally shifting tropical rain belt. In oceanic regions (due to the water column's high heat capacity), the range is small. But on the continents (due to the land's low heat capacity), the seasonal range is much larger. The Amazon rain forest lies within the limits of South America's tropical rain belt. During the periods of Heinrich impacts, this belt was pushed southward. Evidence for this shift comes from Arz's river runoff record (circle) and Edwards's stalagmite growth record (square; see next chapter). This shift is also recorded in Cariaco Basin sediments (triangle).

mainly summer temperatures. Furthermore, in the absence of the supply of ocean heat, the planet's entire northern cap would experience frigid winters! This would explain the pollen record in Alaska and that near our Lamont-Doherty campus. It could even explain why the impacts of the monsoon winds over the Indian Ocean weakened, for these winds respond to the spring

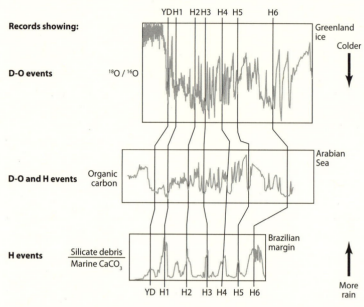

Figure 6-6. Records for the last roughly seventy thousand years from three locales. The point of this figure is to show that some records (such as that for Greenland ice) are dominated by the millennial-duration Dansgaard-Oeschger events, whereas others (such as that for the Brazil margin) are dominated by Heinrich event impacts. Still others (such as that for Arabian Sea sediments) show both. Why the difference? The likely explanation is that the low-latitude locales respond to rainfall changes related to shifts in the position of the tropical rain belt, whereas at high northern latitudes temperature changes are more important.

warming of the Asian continent. Colder conditions would cause the winter snow to linger longer and therefore delay the spring warming and, hence, the onset of the monsoons.

But this left one observation unexplained. Why did a freeze-over of the Northern Hemisphere shift Brazil's rain belts? Enter John Chiang. He realized that were the Northern Hemisphere to preferentially cool, it would push the thermal equator to the south. Furthermore, just as the position of the tropical rainfall belt (see figure 6-5) follows the sun's seasonal course, so also would it respond to a southward shift in the thermal equator. In order to demonstrate that additional Northern Hemisphere sea ice could do the job, Chiang conducted model simulations comparing the positions of the rain belts with and without extended sea ice cover in the northern Atlantic and in the northern Pacific. As predicted, the rain belts were pushed to the south. He found that he could produce a southward shift of the Amazonian rain belt that would not only bring rainfall to dry eastern Brazil and thereby increase the river flow to the sea, but also push the rain belt away from northern Venezuela, depriving it of rain and thereby reducing runoff into the Cariaco Basin (see figure 6-5). Just what the doctor ordered!

Although these findings provide a possible explanation for most of the far-field impacts of the Dansgaard-Oeschger events, the Heinrich events, and the Younger Dryas, there remains the bothersome difference between the pattern and magnitude of impacts, especially the former two. Why do some records show exclusively Heinrich impacts and others exclusively Dansgaard-Oeschger events (see figure 6-6)? Why are Heinrich impacts stronger than D-O impacts? Is the Younger Dryas more akin to a Heinrich event or to a Dansgaard-Oeschger event? Or is the Younger Dryas a special kettle of fish? Although I don't have adequate answers, these are questions to which we must return. But before we do so, there are a few other items on our agenda. The next to be explored is the amazing record kept in caves.

CHAPTER 7

A Confirmation

The mention of rainfall brings to mind monsoons. Although these intense precipitation events occur everywhere in the tropics, we most often associate them with India. I visited there only once. It was during May, when hot and dry weather prevailed. I learned about monsoons as I sat with my friend and host, Devendra Lal, in a circus tent located in a dry river bed. While we waited for the start of a magic show, he explained that in ten or so days' time the tent would be gone, for the onset of the monsoons would flood the now dry bed with a torrent of water. He explained that the dry air that currently flowed down across India from the Tibetan plateau would be replaced by moist air drawn in from the Indian Ocean. The reversal in air flow would be triggered by the late spring heating of the plateau.

Although perhaps less dramatic, a similar sequence of events happens everywhere in the tropics. During winter months, cold and relatively dry continental air drains out to sea. Then when the sun returns and the continents heat up, ocean air is drawn inland and sheds its load of water vapor as torrential rain. One might argue in simpler terms, however, that the seasonal appearance of the Asian monsoons is related to the northward shift of the equatorial rain belt.

As was briefly mentioned in chapter 1, measurements of ^{18}O in stalagmites from caves in China demonstrate that, as might be expected, the strength of monsoon rainfall varies with summer insolation: the greater the amount of sunshine, the stronger the monsoonal impact on the ^{18}O record. Our focus in this chapter will be the exceptions to this relationship, that is, millennial-duration periods of anomalously weak monsoons.

It is important to learn why stalagmites are such a wonderful archive of these anomalies and why it was so late in the game that these records made their debut.

The first thing to be mentioned is that the ingredients for the formation of cave calcite come from the overlying soils. Rainwater percolating through the soil is enriched in CO_2 released by respiration of soil organic material. Through reaction with mineral matter in the soil and underlying limestone, this CO_2-acidified water picks up Ca^{++} and ions of other metals (including uranium). It also picks up additional dissolved inorganic carbon through the dissolution of the limestone through which it passes. Once in the cave, this water drips down from the ceiling and, released from its confinement, its CO_2 begins to escape to the cave air. This loss leads to an increase in the carbonate ion concentration creating supersaturation with respect to $CaCO_3$. As the water runs off the tops of stalagmites, calcite precipitates. Stalagmites prove to be the best of cave paleo-archives for they grow upward in an understandable way.

Back in the late 1960s, Chris Hendy, a chemistry graduate student in New Zealand, did his PhD thesis on the geochemistry of carbon and oxygen isotopes in cave calcite. It was a brilliant piece of research that opened the way to the use of ^{18}O and ^{13}C as paleo proxies. More than two decades would pass, however, before cave studies produced results that made ripples in our community.

The problem was that precise dating proved difficult. The obvious approach was to use radiocarbon. I was the first to publish such an attempt and found that newly deposited calcite had an apparent age of about 1,500 years. Clearly, the reason was that radiocarbon-deficient limestone was dissolved along the water's traverse from soil to stalagmite. The realization that the magnitude of this age offset would vary from cave to cave and perhaps even with time in the same cave discouraged adoption of this dating approach.

Although it was clear from the beginning that the ^{230}Th-^{234}U method could be used on cave deposits, few such measurements were made until the late 1990s. A number of reasons can be given. Those made by the alpha counting method were time-consuming and required relatively large samples. In addition, other applications such as sea level reconstructions based on coral-dating were, at the time, far more attractive. Only when Larry Edwards and his coworkers at Caltech showed during the mid-1980s that atom counting by mass spectrometry was a far better way to do the measurements did it become feasible to study cave formations (i.e., speleothems). But even then, the lure of corals to reconstruct sea level and to extend the calibration of the ^{14}C method[1] took precedence over other applications.

Although a few studies revealed promise, no particular excitement was generated. For example, Israeli scientists produced a ^{230}Th-dated ^{18}O record for their Soreq Cave and showed that

---

[1] Because of changes in the production rate of radiocarbon by cosmic rays and of changes of the distribution of radiocarbon within the ocean, the ^{14}C to C ratios in atmospheric CO_2 and in surface-ocean dissolved inorganic carbons have changed with time, leading to an offset between radiocarbon ages and calendar ages. As shown by the comparison of ^{14}C and ^{230}Th ages obtained on pristine corals, during peak glacial time this offset was a whopping three thousand years!

it nicely tracked that for foraminifera in the near eastern Mediterranean Sea. What opened up the field were measurements made by Larry Edwards and Hai Cheng in cooperation with a number of Chinese collaborators on stalagmites from Hulu Cave. They struck pay dirt big-time. First, they did reconnaissance measurements of uranium and thorium in order to locate stalagmites that would allow them to obtain highly precise ages. They sought material with high uranium to thorium ratios. This required a high uranium to calcium ratio in the cave water and a low content of thorium-bearing mineral detritus. Although due to its very high particle reactivity little dissolved thorium reached the cave, the detritus carried by the water contained thorium (including ^{230}Th). Estimates of the detrital contribution to the total ^{230}Th in any given sample can be made based on the amount of long-lived ^{232}Th present. But as the ^{230}Th to ^{232}Th ratio in soil detritus varies, the correction has a sizable uncertainty. Hence the desire to minimize it by locating a stalagmite as free as possible of mineral detritus.

The other requirement is that the variations in the ^{18}O to ^{16}O ratio in the calcite carry a decipherable and interesting message. At temperate and low latitudes this is a problem because unlike high latitudes, where the ^{18}O to ^{16}O ratio in precipitation varies with temperature, at low latitudes this dependence levels off and other factors become even more important. Of these, the so-called amount effect (see figure 7-1) proves to be dominant. As is the case for air-mass cooling, it leads to rain depleted in ^{18}O. But the reason is quite different.

The vapor that evaporates from the tropical ocean has an initial ^{18}O to ^{16}O ratio about 1 percent lower than that in seawater (because the vapor pressure of $H_2^{16}O$ is about 1 percent higher than that of the heavier $H_2^{18}O$). If the first rain to form carries away only a small fraction of the cloud's vapor, then it will have an ^{18}O to ^{16}O ratio about 1 percent larger than that for the

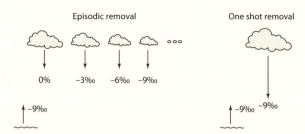

Figure 7-1. Two rainfall scenarios. For precipitation at high latitudes, a so-called Rayleigh condensation dominates. It occurs along the path taken by the air mass. Each time it is cooled, part of its vapor condenses. As the resulting precipitation is enriched in the heavy isotope by one fractionation factor, the remaining vapor becomes ever more depleted and so also does each successive rain (or snow). For precipitation at low latitudes, air masses often lose a major fraction of their moisture in a single event. In the extreme, were it all to be lost during the first rain event, the isotopic composition would be the same as that of the vapor evaporated from the sea surface. For this reason, this type scenario is referred to as the "amount effect." Monsoon rains and thunderstorm rains are examples of situations where the amount effect is important.

vapor and hence its isotopic composition will be close to that in seawater. But if the cloud were to lose a major fraction of its vapor to precipitation during this event, the aggregate composition of the precipitation would have to be close to that in the vapor. One might refer to the "amount effect" instead as the "thunderstorm effect." Thunderstorms are caused by instabilities instigated by heating of the land. The air column becomes gravitationally unstable and the moist ground-level air starts to rise. As it rises, the air cools, causing a portion of its water vapor to condense. The consequent release of heat increases its buoyancy, accelerating the rise of the air parcel. The result is that the plume loses a major fraction of its vapor and hence the resulting rain is anomalously deficient in ^{18}O.

What Edwards and Cheng found was that the Hulu Cave ^{18}O record appeared to be dominated by the amount effect.

Furthermore, the stronger the summer insolation, the stronger the monsoon contribution to the water descending into the cave and hence the more negative the ^{18}O to ^{16}O ratio in the stalagmite calcite. The range they found from summer insolation maximum to summer insolation minimum averaged about four times larger than that for glacial to interglacial change in the ^{18}O to ^{16}O ratio in surface ocean water. Furthermore, the shape of the Hulu record does not resemble the saw-toothed hundred-thousand-year cycle seen in ocean foraminifera and in Antarctic ice. Nor could it be explained by temperature changes in the cave, for four per mil would translate to an unreasonably large 17 or so degree centigrade cooling. Therefore, only one explanation remains: it must be related to the amount effect.

Our interest in this chapter is, however, not in the relationship of the isotopic composition with respect to changing summer insolation but rather in the millennial-duration departures from this dependence. Most striking are the two that punctuate the interval of the last deglaciation (see figure 7-2). One corresponds to the interval immediately following the last of the six Heinrich events, which occurred about 17,500 years ago and was followed by a 3,000-year interval of extensive sea ice cover in the northern Atlantic. It appears from the ^{18}O record in China that the resultant cooling of Eurasia greatly weakened the Chinese monsoons. Then, during the Bølling Allerød, when the conveyor came back into action eliminating the sea ice, as indicated by the ^{18}O record, the monsoon strength returned to that expected based on summer insolation. Finally, a second dip in strength occurred during the time of the Younger Dryas, a time when the conveyor came to a halt and winter sea ice once again covered the northern Atlantic.

One might ask why the monsoons respond so strongly to these millennial-duration sea ice episodes but not to the hundred-thousand-year cycle in land ice. The answer is likely

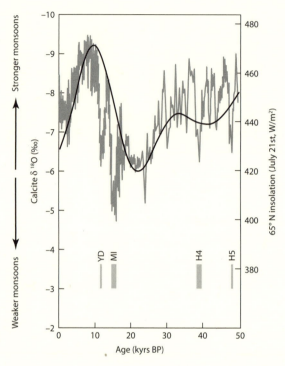

Figure 7-2. As shown in the blowup of the 0- to 60-kyr portion of the Chinese stalagmite 18O record, obtained at the University of Minnesota by Larry Edwards and his group, there are five major departures from the trend expected if the strength of the monsoons slavishly followed the local summer insolation (smooth curve). Edwards refers to them as weak monsoon intervals. They occur during the aftermath of Heinrich events 5, 4, and 2, the Mystery Interval (following Heinrich event 1), and the Younger Dryas (following Heinrich event 0?). As indicated by the growth of Brazilian stalagmites (black bars), these are also times when the tropical rain belt was shifted to the south, bringing rainfall to dry eastern Brazil. The southward shifts were presumably induced by the presence of large extents of winter sea ice in the northern Atlantic. The presence of this ice cuts off the supply to the atmosphere of ocean heat, leading to increased winter cooling of land masses at high northern latitudes. It is this cooling that weakened the monsoons.

that sea ice cuts off the supply of ocean heat to the atmosphere but, of course, land ice does not. Furthermore, the areas that were covered by ice caps are now largely covered with winter snow. Hence, sea ice leads to a larger wintertime cooling than does land ice.

Stalagmites can be used in quite another way: namely, to define times when flow of water through the cave was shut down. For example, during cold periods the water supply might freeze. The Minnesota group came up with another related application. They sampled a cave in a currently dry area of Brazil that lies just outside the limit of the Amazonian rain belt. Using ^{230}Th dating, they showed that its stalagmites grew in short bursts separated by hiatuses representing times when the locale was so arid that the water supply to the cave was shut down. The last two of these growth bursts occurred during the Mystery Interval (aftermath of Heinrich Event 1) and the Younger Dryas. Two earlier ones corresponded in age to the aftermath of Heinrich events 5 and 4 (at 47,000 and 39,000 years before the present, respectively). This finding is consistent with Helge Arz's record of detrital deposition offshore from this dry area. It reinforces the idea that ice cover in the northern Atlantic pushed southward the thermal equator and its associated rain belts.

Also consistent with the Hulu Cave weak monsoon events is the atmospheric methane record (see figure 7-3). It has pronounced lows at the times of the Younger Dryas and the Heinrich events. As much of the planet's methane production occurs in wetlands where water is supplied by monsoon precipitation, its reduction would be expected to decrease the extent of these wetlands and hence the production of methane.

Finally, one other manifestation of the impacts of the millennial events on the Earth's rainfall distribution merits mention. Michael Bender, currently a Princeton professor, showed that

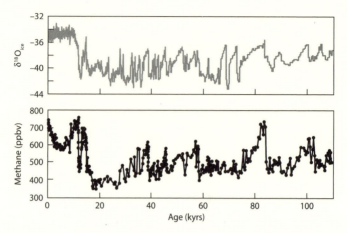

Figure 7-3. Comparison between the record of oxygen isotopes in ice and the record of methane concentrations in air trapped in the ice. As can be seen, during each Dansgaard-Oeschger interstadial (and during the Bølling Allerød), a rise in methane content occurs. The methane measurements were made by Ed Brook of Oregon State University on material from the GISP2 ice core.

the ^{18}O to ^{16}O ratio in the O_2 gas trapped in Antarctic ice underwent cyclic changes. The pattern is remarkably similar to that in Chinese stalagmites (i.e., it has a strong 20,000-year periodicity and a correspondingly weak 100,000-year periodicity). Again this suggests that summer insolation changes, rather than ice volume changes, are responsible.

Before explaining Mike's important discovery, I can't resist reminiscing about his student days here at Lamont. His first scientific discovery was made while he was a summer intern, when he demonstrated that the manganese nodules that pepper much of the abyssal sea floor grow at the incredibly slow rate of millimeters per million years! When a year later he was admitted as a Columbia graduate student, he decided that he needed to learn some geology and that this was best done in Israel. A year later, he returned to New York endowed not only with geology, but

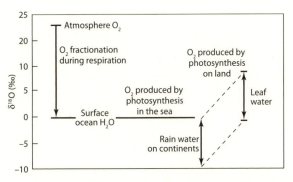

Figure 7-4. While no oxygen-isotope fractionation occurs during photosynthesis, it does during respiration. O_2 molecules with two [16]O atoms are consumed with about 2 percent higher probability than those containing one [18]O and one [16]O atom. Because of this, today's atmospheric O_2 has a 23 per mil higher $\delta^{18}O$ than surface seawater. Complicating the situation, on land the water from which O_2 is produced has a slightly different $\delta^{18}O$ than seawater. Furthermore, this offset changes with climate. These changes reflect the fact that the source of water for land plants is rain. Countering the deficiency of [18]O in rain is fractionation during evaporation from the leaf surface, which enriches leaf water in [18]O. The result is that the so-called Dole effect changes with climate. These changes are recorded in the O_2 trapped in ice-core bubbles.

with the independent spirit characteristic of that nation. He took up big-time opposition to the Vietnam War. I well remember leaving a full faculty session (held in Columbia's chapel) dealing with how to cope with the ongoing campus-wide protest, only to find Mike sitting cross-legged on the sidewalk with a serious look on his face and his fingers giving us the peace symbol. Although often distracted, Mike went on to do a superb thesis and thereafter a series of successful research projects which earned him membership in the National Academy of Sciences.

To understand what Mike's discovery is trying to tell us requires delving into what is referred to as the Dole effect (see figure 7-4). Although no significant isotopic separation occurs during photosynthesis (i.e., the isotopic composition of the O_2 produced is close to that in the plant's leaf water), this is not the case however for respiration. Light $^{16}O\ ^{16}O$ molecules are consumed in slight preference to heavy $^{16}O\ ^{18}O$ molecules. Because of this, the atmosphere's oxygen gas has built up an ^{18}O to ^{16}O ratio a little more than 2 percent larger than that in ocean water oxygen. As global photosynthesis produces an amount of O_2 equal to that contained in the atmosphere in about a millenium, on this timescale the oxygen isotope composition reflects the "details" of the O_2 isotopic cycle.

I say "details" because there are several reasons why the isotopic composition of oxygen gas changes with time. First, as we have seen, the isotopic composition of seawater changes as the ice sheets wax and wane. Furthermore, photosynthesis on land produces O_2 from leaf water whose oxygen-isotope composition reflects that in continental precipitation (modified by fractionation during evaporation from leaf surfaces; see figure 7-4). Finally, the fractionation during respiration has its own small variability.

The strong twenty-thousand-year cycles in the ^{18}O to ^{16}O ratio in ice-core oxygen gas tell us that influences other than the isotopic composition of seawater are involved (see figure 7-5). Although to date no agreement exists as to what combination of these effects generate the observed twenty-thousand-year cycle, it is clear that it must reflect, in one way or another, the changing isotopic composition of the O_2 produced on land. Hence it must reflect the planet's hydrology.

With this in mind, Jeff Severinghaus took this proxy one step further. He did more accurate measurements of the ^{18}O to ^{16}O ratio in O_2 bubbles in Greenland ice. What he found was that, during the course of each millennial event, the ratio underwent a gradual change. At the times of each abrupt warming and each abrupt cooling (as marked by sudden shifts in methane

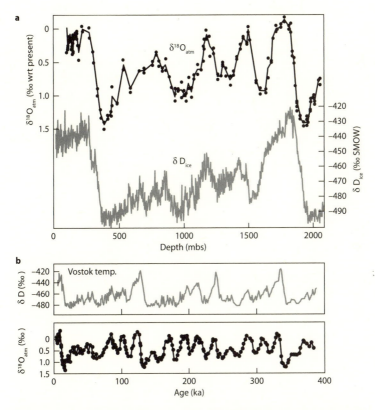

Figure 7-5. (a) Comparison between Michael Bender's oxygen-isotope record for the O_2 trapped in the bubbles of Antarctic's Vostok ice core and Jean Jouzel's hydrogen-isotope record for the ice itself. As can be seen, the former record has a far stronger twenty-kyr component than the latter. The changes in the $\delta^{18}O$ in O_2 primarily reflect changes with climate in the isotopic composition of leaf water. (b) A comparison of the $\delta^{18}O$ in O_2 with Vostok air temperature over the last 380 kyrs.

concentration in the same air bubbles), however, an abrupt change in the slope of the isotope ratio trend occurred (see figure 7-6). The difference in response between the methane concentration and the O_2 isotope ratio to these abrupt climate

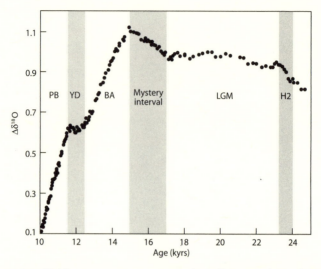

Figure 7-6. Highly accurate measurements by Jeff Severinghaus of the ^{18}O to ^{16}O ratio in O_2 from bubbles in the Siple Dome Antarctica ice core. Note the change in slope that takes place at the boundaries between events. Each slope change reflects the onset of the drift of the isotopic composition of the atmosphere's O_2 toward a new steady-state value. The zero on the $\Delta\delta^{18}O$ scale is the isotopic composition of today's O_2.

changes has to do with their respective atmospheric residence times. Methane is replaced on a decadal timescale. Oxygen is replaced on a millennial timescale. This being the case, the methane concentration changes abruptly at each transition but the ^{18}O to ^{16}O ratio in O_2 starts to drift slowly toward a new steady-state value. Then when the next transition occurs, the slope of the drift shifts to a different value.

So it all fits together nicely. The fresh water cap created by the melting of each Heinrich armada shuts down the conveyor and this allows winter sea ice to form over much of the northern Atlantic. The presence of this sea ice cools Eurasia, thereby weakening the monsoons. It also pushes the tropical rain belt to the south. Of interest is that the response to the Younger Dryas event was very much like that to the Heinrich events. In the next chapter we will look in detail at the hypotheses regarding the origin of the Younger Dryas. Could it have been a seventh Heinrich event?

CHAPTER 8

The Last Hurrah

As already mentioned, for many years I was convinced that the Younger Dryas was a freak event that resulted from a one-time catastrophic flood of water into the northern Atlantic. I even joked by saying that God had placed it at the close of the last glaciation as a warning of what might happen if we added too much CO_2 to the atmosphere. During the last few years, however, new information has convinced me that this is not the case; rather, the Younger Dryas was very likely an integral part of the sequence of events associated with glacial terminations.

Before discussing what changed my mind, let me review the evidence that for many years sent me in what I now think was likely the wrong direction. First, as part of the publication of the material in my PhD thesis, an article titled "Evidence for an Abrupt Change in Climate Eleven-thousand Years Ago" came out in 1960. My coauthors were Maurice Ewing, the originator and first director of the Lamont Laboratory, and Bruce Heezen, my choice as Lamont's greatest scientist. Impressive company! The article summarized results from deep-sea sediments in the Atlantic, from shorelines of the closed-basin Lake Lahontan, and from sediments from the Mississippi's delta. All showed what appeared to be a unidirectional sudden termination of glacial conditions. A few years later, I named these abrupt endings Terminations. So when it became clear to me that the last of these Terminations more likely occurred at the onset of the Bølling Allerød, I was predisposed to think of the Younger Dryas as a freak mishap tagged onto the end of the last glacial period.

This idea was reinforced by articles I read concerning a sudden release of water stored in proglacial Lake Agassiz, which formed during the retreat of the Laurentide ice sheet. The evidence came from two sources (see figure 8-1). The first was from the lake itself indicating that its level had undergone a sudden thirty-meter drop. Radiocarbon dates on wood lodged in the shoreline deposits marking its low stand placed the timing of this drop within the dating error of the onset of the Younger Dryas. The second was a detailed ^{18}O record from high deposition–rate sediment cores from the Gulf of Mexico, which revealed a sharp rise in ^{18}O to ^{16}O ratio dated by radiocarbon to have occurred close to the time of the onset of the Younger Dryas. Together these two observations appeared to be telling us that at this time, a portion of the retreating Laurentide ice sheet that formed the lake's northern shoreline gave way, released a huge amount of water, and also brought to a halt the flow through the outlet that had routed the overflow water into the drainage of the Mississippi River. Because the ice-sheet meltwater that had passed through Lake Agassiz into the Mississippi River was depleted in ^{18}O relative to local precipitation, this diversion neatly accounted for the ^{18}O increase observed in Gulf of Mexico planktic foraminifera when the southward outflow was diverted.

This seductive scenario, however, had one flaw, which proved fatal. No convincing geomorphic evidence had been reported along the path taken by the floodwaters across the region now occupied by the Great Lakes and out the St. Lawrence Valley into the northern Atlantic. Being more a geochemist than a

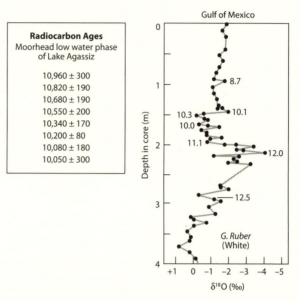

Figure 8-1. Summary of radiocarbon ages on wood samples from the Moorhead low stand shoreline of Lake Agassiz. For comparison, the radiocarbon age for the onset of the Younger Dryas is 10,900 ± 50 years. Also shown is the ^{18}O to ^{16}O record for a core from the Gulf of Mexico's Orca Basin. The more negative values for the Bølling Allerød interval (12.5 to 11.0 radiocarbon kyrs) are interpreted as reflecting the overflow of ice-sheet meltwater via Lake Agassiz. The increase that occurred close to 11.1 radiocarbon kyrs ago is interpreted as the result of the diversion of Agassiz overflow to an alternate route. As the radiocarbon dates are on foraminifera shells, a reservoir correction of 0.4 kyrs has been applied.

geologist, I was able to put this flaw aside until I learned about Ouimet Canyon just north of the western end of Lake Superior (see figure 8-2). It is a one-hundred-meter-wide and one-hundred-meter-deep, two-kilometer-long channel cut through a diabase sill much like the one on which my lab sits. This channel is now dry and has no apparent reason for being. "Aha!" I said, "It must be the product of the Agassiz flood." But when I

Figure 8-2. Dry Ouimet Canyon, located northeast of Thunder Bay, Ontario, Canada. It was cut through an Archean-age diabase sill by a meltwater flood that is thought to have occurred sometime during the last period of deglaciation. Unfortunately, no means of determining its exact age has been found. Pictures by Gary Comer, left, and Jim Teller, right.

queried Jim Teller of the University of Manitoba, he said, "No, no, Wally. Ouimet was created by a younger event. Its locale was still under ice during the Younger Dryas. The Agassiz flood passed to the south through what is now the city of Thunder Bay." He went on to tell me about a series of younger channels extending one hundred or so miles to the north of Ouimet supposedly cut by successive floods generated as the Laurentide ice sheet's front continued its northward retreat.

At this point, I was hit by the lack of geomorphic evidence for the flood presumed to have triggered the Younger Dryas. If subsequent floods described by Teller created gashes in the landscape but were not big enough to shut down the conveyor,

Figure 8-3. Boulder field to west of the northern end of Lake Nipigon in Ontario, Canada. The boulders are granitic and range in diameter up to 1.5 meters. As can be seen, the field contains a major "ripple," indicating that the boulders were emplaced by a very strong current. As determined by Meredith Kelly, the [10]Be age of the boulders is close to eleven thousand calendar years. Picture by Tom Lowell.

how could it be that no gash was found recording the flood I had been touting as the trigger for the conveyor shutdown that initiated the Younger Dryas?

When I expressed these concerns to my friend Gary Comer, he suggested that we go and have a look to see what evidence lay along Teller's proposed route. So a group of three experts—Jim Teller, George Denton and Tom Lowell—joined Comer in his Caravan airplane and flew west from Thunder Bay in search of the channels and boulder fields (see figure 8-3) similar to those found to the north of Ouimet. Nary a hint of flood activity was to be seen.

A somewhat chagrined Teller reminded us of a huge channel in the vicinity of Fort McMurray that had been cut by floodwaters streaming northward to the Arctic Ocean. Although radiocarbon dates suggested that it was created a couple of thousand years after the Younger Dryas, perhaps the channel was cut by an earlier flood and reoccupied by a second one. Although this seemed unlikely, Comer once again offered his Caravan, this time equipped for water landings. But alas, once again we were foiled. No evidence of an earlier flood was to be found.

Could it be that a catastrophic release of meltwater stored in Agassiz was not the trigger for the Younger Dryas after all? What about the shoreline evidence in Agassiz itself? Tim Fisher of the University of Toledo pointed out that the timing was based on the radiocarbon dating of a single wood sample. Maybe, he suggested, the wood was reworked and the lake dropped at a later time and more slowly—that is, no catastrophic flood. If so, what about the evidence from the Gulf of Mexico? I had always been aware that an alternative explanation existed: namely, that the large production of meltwater that characterized the Bølling Allerød warm period was largely shut down at the onset of the Younger Dryas cold snap. This would have altered the blend of local precipitation and ^{18}O-depleted glacial meltwater coming down the Mississippi. In this case, a sudden rise in ^{18}O would reflect a *reduction* rather than a *diversion* of meltwater.

About the time these unsettling thoughts began to rattle around in my brain, my friend George Denton astounded me by making a clear and convincing argument that the Younger Dryas could not, as I had thought, have been a freak coincidence tacked onto the end of the last glacial period. His argument ran as follows. Looking from the other end of the world, the Bølling Allerød could not have marked the beginning of the present interglaciation because only half the warming of Antarctica had taken place and CO_2 had risen only half-way to the content achieved during previous interglacials. As both the Antarctic temperature and the atmospheric CO_2 had plateaued, something had to jar them loose and send them on their way to the full interglacial condition!

A week or two later, Larry Edwards delivered the knockout blow. His stalagmite records from Chinese caves suggested to him that Termination III[1] was interrupted by a pair of weak monsoon events separated by a recovery to the insolation-dictated monsoon strength. This tripartite series of events reminded him of the Mystery Interval–Bølling Allerød–Younger Dryas sequence that punctuated the last deglaciation.

So, if it wasn't a flood that halted the conveyor and thereby triggered the Younger Dryas, what was it that terminated the pause in the Antarctic warming and the pause in the CO_2 rise? John Andrews and others have proposed that the Heinrich event 1 that occurred close to the beginning of the Mystery Interval was not, after all, the last in the series of ice armadas. There was a subsequent one, which he termed Heinrich event 0. Evidence for it comes from a layer of glacial detritus in sediment cores from close to the mouth of the Hudson Straits. Although not well dated, its age is broadly consistent with that for the onset of the Younger Dryas. Even though unlike the other six Heinrich events, it is not recorded in sediments farther away from the Hudson Straits; the far-field Younger Dryas impacts do closely mirror those associated with the Heinrich events. Most intriguing of these is the burst of stalagmite growth seen in the Brazilian cave.

Early on, in an attempt to explain the multiple Dansgaard-Oeschger events, I proposed a salt oscillator. My idea was that, if the flow of salt out of the Atlantic got out of sync with the export of water vapor, the system might jump back and forth between a conveyor-on and a conveyor-off mode of operation. If the conveyor were to stop, then the buildup of salt due to water

---

[1] Termination I took place about 12 kyrs ago, Termination II about 130 kyrs ago, and Termination III about 250 kyrs ago.

vapor export would likely exceed salt export from the Atlantic. Under this circumstance, the salt content and hence the density of Atlantic water would increase and eventually become great enough to kick the conveyor back into action. If, under this new condition, salt was exported faster than it was being enriched by water vapor export, then the density of the Atlantic water would coast back down until the conveyor could no longer function. Certainly this was an overly simplistic idea, but perhaps it had the seeds of a viable mechanism.

Once alerted, ocean dynamicists turned their attention to this subject. Some of their models suggested yet another explanation. These models showed that in its present state, the Atlantic prefers to be in its conveyor-on mode of operation. But under glacial conditions, the models assumed alternate modes of operation. If so, then there must have existed what dynamicists refer to as bifurcation points at which the circulation would switch from one of these modes of operation to another (see figure 8-4). If the models' conditions were set to lie close to such a bifurcation point, then a small perturbation (rather than a large pulse of meltwater) would be capable of pushing the system over the brink from one mode of circulation to another. If the real ocean operates like these models, then there would be no need to search for a powerful trigger event. As the modeler has at his disposal many "knobs to turn," I have never been convinced that the real ocean necessarily performs in this way. But because my credentials in the field of ocean modeling are lacking, my opinion has little value.

I can't end this chapter without mentioning what I consider to be a rather outrageous new hypothesis regarding the origin of the Younger Dryas. It was published in 2007 in the prestigious *Proceedings of the National Academy of Sciences*. Among the twenty-four authors was the same Jim Kennett who had produced the marvelous Santa Barbara Basin record. In the article,

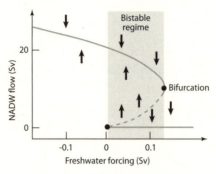

Figure 8-4. Bifurcation point in a model simulation of the circulation in the Atlantic Ocean conducted by Stefan Rahmstorf. As the amount of fresh water added to the northern Atlantic by precipitation and river runoff increases, the strength of the models overturning circulation decreases until a point is reached where the conveyor shuts down and the system jumps to its alternative mode of operation. In order to reinitiate the conveyor, the freshwater forcing must be significantly reduced. One Sverdrup (Sv) corresponds to a flux of one million cubic meters of fresh water per second.

the authors propose that the Younger Dryas was triggered by what scientists term an "air blast." Such events occur when a small piece of a comet or an asteroid smashes into the atmosphere and burns up. The most famous such event occurred in 1908 in the Tunguska region of Siberia, when a blast flattened two thousand square kilometers of forest (see figure 8-5). The authors of the PNAS article list an impressive array of evidence in support of such a blast. It comes from sites stretching from Belgium to California and from Greenland to Arizona. The evidence includes buckyballs (soccer ball–shaped carbon molecules thought to be produced in supernovas), nanodiamonds, the metal irridium, spherules purported to be produced in intense fires, and peaks of nitrate and ammonia in the Greenland ice core supposedly generated by the high-temperature

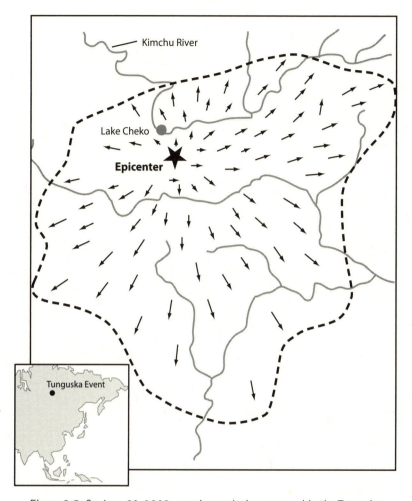

Figure 8-5. On June 30, 1908, a major explosion occurred in the Tunguska region of Siberia. It caused destruction in a two-thousand-square-kilometer area of Taiga. Trees were flattened. Fragments of the impacting body have never been found, leading to a consensus that the impacting object was vaporized in the overlying atmosphere. Hence it is referred to as an air blast. The figure was prepared by Luca Gasperini and Enrico Bonatti.

reaction between atmospheric N_2 and O_2. They also list radio-carbon dates showing that the horizons in which these entities are found correspond in age to the onset of the Younger Dryas. Furthermore, they link this event to the disappearance of the early Americans who manufactured the beautiful Clovis arrow-heads and also to the extinction of mastodons, saber-toothed tigers, and other New World mega fauna. At first look, this evidence appears to be overwhelming, and if the authors have indeed found what they claim to have found, one would have to accept that they have made a stupendous discovery. But as cross-checks on and evaluations of these impact proxies roll in, one by one they tumble. By my last count, only nanodiamonds remained standing, but as the validity of this remaining proxy rests on very shaky footing, I suspect that this bizarre scenario will soon bite the dust.

In any case, the Younger Dryas event is likely to occupy the attention of paleoclimatologists, archeologists, and modelers for many years to come!

CHAPTER 9

Holocene Wobbles

During the eleven thousand years that have elapsed since the end of the Younger Dryas, only two events that might be classified as conveyor shutdowns have interrupted the relative quietude of the present interglacial period. One, the 8.2-kyr event, has received wide attention, in part because it appears to have been triggered by a flood and in part because it shows up in far-flung records (see figure 9-1). The source of the fresh water is thought to have been a proglacial lake that was situated to the south of what is now the Hudson Bay. As did Lake Agassiz, this lake underwent a large drop in level. The proposal is that the water released during this drop forced its way beneath the small remnant of the Laurentide ice sheet and passed through what is now the Hudson Bay and then out the Hudson Straits into the Atlantic Ocean. This water is presumed to have created a low-salinity cap, which temporarily shut down the conveyor. I say "temporarily" because, unlike the Younger Dryas, which lasted 1,350 years, the 8.2-kyr event lasted only about 80 years. Modelers attribute the difference in duration to what they refer to as the preferred conveyor-on mode of ocean operation. Despite its short duration, the 8.2-kyr event caused a drop in the methane

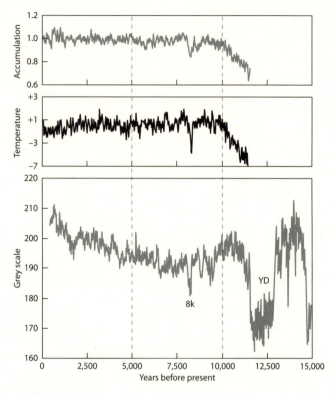

Figure 9-1. In the upper panel are shown the accumulation rate (based on annual layer thickness) and temperature (based on oxygen isotope ratios) for the Holocene portion of the GISP2 Greenland ice-core record. The brief 8.2-kyr event stands out. In the lower panel is shown the gray-scale record from the Cariaco Basin for the last 15 kyrs. Again, the brief 8.2-kyr event stands out. Data courtesy of Richard Alley.

content of the atmosphere and a shift in the slope of the ^{18}O to ^{16}O trend in O_2. It also weakened the Asian monsoons.

The other candidate for a conveyor shutdown is the preboreal event that occurred very early in the Holocene. Although it receives far less attention than the 8.2-kyr event, it also resulted in a drop in atmospheric methane content and a shift in the slope of the O_2 isotope ratio trend. It is possible that a single flood responsible for the creation of the entire series of channels and boulder fields to the north of Ouimet Canyon triggered this event. Meredith Kelly, of Dartmouth College, has obtained [10]Be dates on several of the granitic boulders from each of the two large boulder fields. The ages cluster around 10,800 years. Each field is made up of thousands of boulders with sizes ranging from one-half to two meters in diameter. The boulders in the northern field form a mega ripple, suggesting that they were emplaced by a deluge (see figure 8-3).

Although the past eight thousand years have been free of conveyor shutdowns, there have been small climate changes that might be attributed to fluctuations in the strength of the conveyor. But as considerable debate has arisen regarding their magnitude and geographic reach, their cause remains in question.

Certainly the most interesting but also the most controversial evidence is Gerard Bond's red-stained grain record (see figure 9-2). As an outgrowth of his research on Heinrich events, Bond made detailed studies of ice-rafted grains in ambient northern Atlantic sediments. What he found was rather amazing. He noted that some of the quartz grains had red stains (presumably hematite).[1] Others did not. In samples from a northeastern Atlantic sediment core, he noted that the proportion of red-stained grains fluctuated back and forth in a cyclic manner. Their abundance ranged from lows of 3 to 6 percent and highs of 15 to 18 percent of the total grains. He tracked these cycles through the entire Holocene and from there well back into the last glacial period. Based on radiocarbon dates, he showed that the time period encompassed by each set of ten cycles was

---

[1] When iron in its +2 valence state encounters atmospheric O_2, it is oxidized to its +3 valence state. In this form, it is highly insoluble and thus is rapidly deposited as hematite coatings. The quartz grains in red sandstones receive their color in this way.

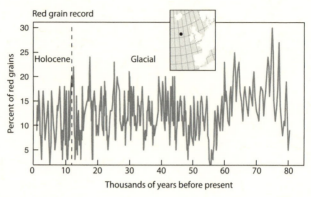

Figure 9-2. Percentage of hematite-stained grains in the ice-rafted debris from a northern Atlantic sediment core for the last 80,000 years, as obtained by Gerard Bond of Lamont-Doherty. On the average, the maxima are spaced at 1,500-year intervals. Although the minima are not as low and the maxima a bit higher, the record for much of the last glacial period is remarkably similar to that for the Holocene. The core site is shown on the inset map.

uncannily close to 14,700 years. Bond initially postulated that the abundance of red grains varied with ocean temperature. His reasoning was that the red grains likely originated in multiyear sea ice formed along shelves surrounding islands of Canada's Arctic archipelago, a region rich in red sandstones. The trick was to float the ice all the way to the site of his core before it melted and dropped its red grains. Therefore Bond postulated that the higher the percentage of hematite-coated grains, the colder the surface of the northern Atlantic Ocean.

Intrigued by the seeming periodic character of his record, Bond decided to compare it with that for the production rate of radiocarbon. The idea was that fluctuations in the production rate of this cosmogenic radioisotope were related to the strength of the so-called heliomagnetic field generated by ions streaming out from the sun's dark spots. Over the past thirty years, the sun's

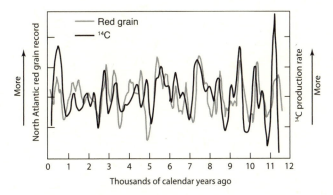

Figure 9-3. Gerard Bond compared his red-grain record with that of the production rate of radiocarbon atoms in the atmosphere. In order to focus on variations that occurred on a millennial timescale, he removed the long-term trend from the record. As can be seen, the match between the variations in cosmic ray production and that in the percentage of red-stained grains is reasonably good. This match led him to conclude that the red-grain cycles were paced by the sun.

irradiance has been precisely measured by specially equipped satellites and, as was suspected, the irradiance has varied in accord with the well-known eleven-year sunspot cycle. The more spots, the higher the solar irradiance. More spots corresponded to a larger emission of ions and hence to a larger magnetic shielding of incoming cosmic rays. The result is a lower production rate of ^{14}C. Bond's strategy was to use the radiocarbon record as a means of determining whether his red-grain abundance varied with solar irradiance. With smaller irradiance, the Earth would cool and the abundance of red grains would be larger.

To everyone's amazement, when plotted on top of one another, the features in the radiocarbon production-rate record (reconstructed from radiocarbon measurements on dendrochronologically dated tree wood) showed a remarkable resemblance to the red-grain abundance record (see figure 9-3). In response to critics who proposed that the radiocarbon record might be strongly influenced by changes in the rate of deep ocean ventilation, Bond pointed to the ^{10}Be record in ice cores.[2] It agreed with that of ^{14}C. Any changes in the rain of cosmic rays onto the Earth's atmosphere would change ^{14}C and ^{10}Be production to the same extent. Since it is difficult to see why changes in ocean circulation would influence the rain of ^{10}Be onto ice caps, the similarity in the two records points to the sun rather than to the ocean.

But there remains a big problem with this linkage. The measurements by satellite reveal that the magnitude of the irradiance change from sunspot maxima to sunspot minima is only about one part in 1,300. As such, the irradiance change should produce temperature changes of only about 0.1°C. Yet the temperature changes associated with Bond's cycles appear to be as large as 1°C. Furthermore, during the so-called Maunder minimum, a seventy-year time interval in the seventeenth century when sunspots disappeared altogether (see figure 9-4), the Earth did not get significantly colder than it was during the previous century or the subsequent century. Hence it is difficult to see how the tiny irradiance changes could lead to such large impacts.

Tragically, Gerard Bond succumbed to cancer before he was able to complete either his attempt to reproduce the red-grain record or to validate his ideas regarding their origin. Only recently has Ben Flower of the University of South Florida picked up this line of research where Bond left off.

One of Bond's red-grain cycles appears prominently in the historic climate record: namely, the Medieval Warm (800 to 1300 AD)–Little Ice Age (1300 to 1850 AD) oscillation. This

---

[2] The ^{10}Be atoms produced by cosmic ray impacts quickly become attached to atmospheric aerosols, which, in turn, are promptly scavenged by raindrops and snowflakes and carried to the Earth's surface.

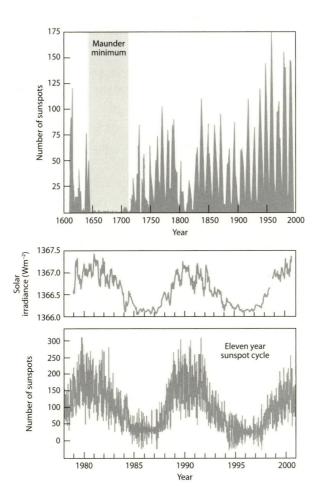

Figure 9-4. Galileo discovered sunspots in 1604 AD. Since then, their number has been monitored (from telescopes), revealing a strong eleven-year cycle. Only during the period from 1645 to 1715 were these dark spots largely absent. Only since 1978 have measurements of sufficient accuracy been conducted (by satellites) to assess changes in the sun's irradiance. During the course of two eleven-year cycles, the irradiance varied by a tiny amount (one part in 1,300). Unless the change was considerably larger during times such as the Maunder Minimum when sunspot activity was much reduced, it is difficult to see how these irradiance changes could perturb the Earth's climate.

temperature fluctuation is best documented at high northern latitudes and it is best known because it appears to have played a key role in the history of the Viking settlements in Greenland. When those colonists arrived during the tenth century they were able to grow enough grain to comfortably feed themselves and their livestock. Carbon isotope measurements on radio-carbon-dated bones from Viking burials suggest that initially 80 percent of their nourishment came from land and only 20 percent from the sea. In the thirteenth century, toward the end of their occupation of Greenland, these measurements suggest that the situation had reversed: only 20 percent of their nour-ishment was derived from the land and 80 percent from the sea. All indications are that a cooling associated with the onset of the Little Ice Age reduced the already short growth season to the point where their grain crops failed. Combined with dete-riorating conditions at sea, which made communication with Scandinavia difficult, the lack of an adequate food supply ap-pears to have snuffed out the colony.

Although the climatic evidence from Greenland is perhaps not as robust as one might prefer, that from the extent of glaciers in Alaska, Iceland, Greenland, Scandinavia, and Switzerland-Austria is overwhelming (see figure 9-5). In each of these places, the advances of mountain glaciers during the Little Ice Age were the largest since the Younger Dryas. As there is a consensus among geophysicists that temperature is the dominant control on snowline elevation[3] and hence on glacial extent, this strongly suggests that, at least at these places, the Little Ice Age was the coldest time interval in the Holocene.

Thanks to efforts by Joerg Schaefer and his group at Lamont-Doherty, a great improvement has been made in the application

---

[3] The snowline is the boundary between the upper portion of the glacier, where snowfall exceeds ablation, and the lower portion, where ablation exceeds snowfall.

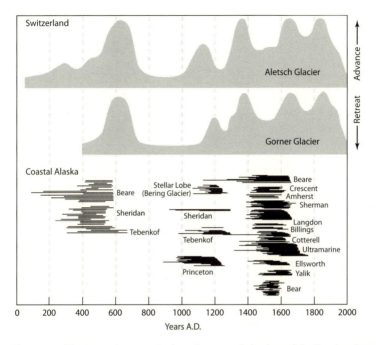

Figure 9-5. The upper two panels show the records for two of Switzerland's major glaciers, the Aletsch and the Gorner, as reconstructed by Holzhauser based on dendrochronologic and radiocarbon data. The lower panel shows times when advancing glaciers killed Alaskan trees, as reconstructed by Wiles. The growth intervals of these trees are shown by the horizontal bars. This diagram was prepared by George Denton.

of the [10]Be-dating method as applied to quartz grains in boulders left behind on Holocene-age moraines. Schaefer abandoned the use of commercial reagent-grade beryllium as a chemical carrier because it contained too much [10]Be. Instead, they extracted beryllium from ancient beryl minerals free of this radioisotope. Furthermore, they had their samples analyzed at Livermore National Laboratory, which operates an accelerator mass spectrometer with the highest beam intensity of any currently in use. In this way, a larger fraction of the ^{10}Be atoms in the sample could be detected. Combined with a very low carrier correction, this enabled Schaefer's group to obtain ages on historically dated moraines with the incredible precision of ±15 years. The fact that the ages they obtained are consistent with the historical ones indicates that the glaciers pluck boulders that have never before been exposed to significant cosmic ray bombardment.

As this book was being written, Schaefer's ^{10}Be ages were rolling in from Holocene moraines in both the European Alps and the New Zealand Alps. One aspect of the records in the two hemispheres stands out. In the north, the largest Holocene glacial advance was during the Little Ice Age. But in the south, the extent has steadily decreased since the early Holocene. This is consistent with the antiphasing of the twenty-thousand-year component of summer insolation. In the south it has been increasing during the course of the Holocene, and in the north it has been decreasing.

To date no clear picture has emerged regarding the phasing of glacial advances in north and south. The preliminary results suggest that, rather than being either in phase with one another or antiphased with one another, the phasing appears to be random. I was disappointed by this result because it appears to have put to rest a hunch that the cause of the millennial-duration fluctuations was some sort of seesawing in the relative strength of deep-water formation at the two polar regions. Were this the case, one might have expected that cold phases in the Northern Hemisphere would correspond to warm phases in the Southern Hemisphere and vice versa. But at this point Joerg Schaefer's ^{10}Be results do not support this idea.

Nevertheless, I still feel that the ocean plays an important role in the millennial cycles, just as it does in the El Niño–La Niña cycle and in the other decadal and multidecadal climate oscillations. Compared to the atmosphere, the ocean is an incredibly large reservoir of heat. Hence small changes in its operation can produce significant fluctuations in the temperature of the overlying atmosphere.

Another aspect of ocean operation that has long puzzled me is the near-equality in the amounts of deep water produced in the northern Atlantic and in the Southern Ocean. As I can think of no obvious feedback mechanism that would maintain this equality, it wouldn't surprise me if it has changed over the course of the Holocene. Indeed, there are two lines of evidence that such a change did occur over the course of the Medieval Warm–Little Ice Age oscillation. Jean Lynch-Stieglitz of Georgia Tech has made use of the relationship between the oxygen isotope ratios in benthic foraminifera shells and the density of bottom water in which they live to reconstruct past changes in the strength of the conveyor. In particular, she has used thermocline-depth ocean sediment cores taken on either side of the Florida Straits to reconstruct the strength of the Gulf Stream (i.e., the conveyor's upper limb). She first studied glacial-age sediments and demonstrated that the flow of this mighty current was reduced by a factor of two compared to today's. She attributed this to a severe weakening of the conveyor. More recently, she has looked at records for the past few thousand years and has concluded that during the Little Ice Age the Gulf Stream weakened by about 10 percent. As about half of the water transported by this current feeds deep-water formation (the other half forms the temperate North Atlantic wind-driven gyre), this suggests a reduction in conveyor strength of up to 20 percent.

The other observation indicating a change in conveyor operation associated with the Medieval Warm–Little Ice Age oscillation was made by Lloyd Keigwin of Woods Hole Oceanographic Institution. It involves radiocarbon measurements on benthic foraminifera shells from a rapidly accumulating sediment at a depth of 4.8 km to the northeast of Bermuda (see

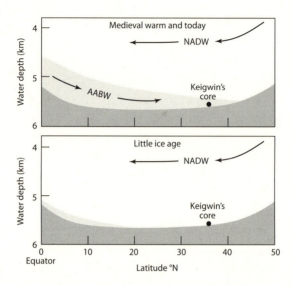

Figure 9-6. Penetration of Antarctic bottom water (AABW) into the western part of the northern Atlantic. Lloyd Keigwin of Woods Hole Oceanographic Institution used radiocarbon measurements on benthic foraminifera shells to demonstrate that during the Little Ice Age this tongue was pushed back by the North Atlantic deep water (NADW). Recently, the AABW tongue has pushed its way back into the northern Atlantic and now reoccupies the "territory" it held during the time of the Medieval Warm.

figure 9-6). He showed that those shells deposited during the Medieval Warm and during the twentieth century contained the ^{14}C-depleted carbon characteristic of the thin tongue of Antarctic bottom water that underrides the North Atlantic deep water mass. By contrast, those shells formed during the Little Ice Age had a 5 percent higher ^{14}C content characteristic of North Atlantic deep water. This demonstrates that during the Little Ice Age, the tongue of Antarctic water was not able to penetrate as far up the Atlantic as it does today.

Although taken at face value the Lynch-Stieglitz observation suggests that the conveyor was weaker during the Little Ice

Age and the Keigwin observation suggests that it was stronger, one could argue that changes in both end members occurred, causing the tongue of Antarctic origin to retreat and the flux of Atlantic water to wane. But the important point is, rather, that these observations demonstrate that the Little Ice Age cooling event involved the ocean. Of course, they do not tell us whether this involvement was active or passive. In other words, did the change in ocean operation drive the temperature change or did the temperature change drive the change in ocean operation? More simply put, was the climate change perturbed from above by the sun or from below by a change in ocean operation?

So, although the ocean's conveyor did not experience any major flip flops during the past eight thousand years, it appears to have undergone minor fluctuations.

CHAPTER 10

The Anthropocene

Paul Crutzen, a Nobel laureate, has proposed that the geologic time interval known as the Holocene came to an end at the onset of the Industrial Revolution and that we have entered what he terms the Anthropocene. His logic is that at this point man's influence on the atmosphere and ocean began to compete with nature's. Of these impacts, that resulting from the buildup of fossil fuel CO_2 is likely to have the greatest impact. The late Roger Revelle referred to the onging CO_2 buildup as "man's greatest geophysical experiment."

From the conveyor concept's very inception, the question arose as to whether the ongoing increase in greenhouse gases might bring it to a halt. Or could it be that the Earth's climate system has other so-called tipping points? Based on the climate system's erratic behavior during glacial time, in my lectures and writings I frequently refer to the Earth's climate system as an "angry beast." In fact, strung across the wall outside my office in the new Comer Geochemistry Laboratory is a sixteen-foot-long pink and blue cloth snake bearing a sign that reads, "I am the climate beast and I am angry."

Early on I realized that if such a shutdown were to occur, it would likely serve to cool only a small region of an already overly warm planet. For, as we have seen, the far-field impacts of the conveyor shutdowns of glacial time were the result of sea ice formation. In a warmer world, no sea ice would be created by a conveyor shutdown. Thought of this way, it would be a disruption rather than a disaster. There is one aspect of the paleo record that did, however, concern me. It was the short-term oscillations seen in the electrical conductivity of Greenland ice at the times of conveyor startups and shutdowns (see figure 10-1). Over a period of a few decades, the system appeared to jump back and forth between one state of operation and another. My concern was that if these stutters were to accompany a future

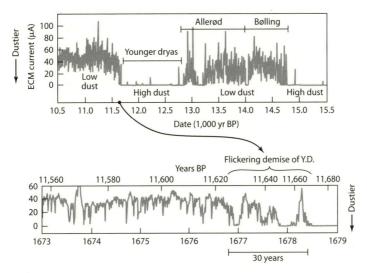

Figure 10-1. Example of the flickers in electrical conductivity (as measured by Ken Taylor of Nevada's Desert Research Institute) that punctuate the abrupt transitions from one mode of ocean-atmosphere operation to another. As shown in the blowup at the end of the Younger Dryas cold snap, the conductivity jumped from its near zero cold-state value to its high warm-state value in a single year and then several years later plunged back to near zero.

conveyor shutdown, they might seriously affect the world's food supply. Each jump would come with so little warning that farmers would not be able to react fast enough to compensate for the change in growth conditions. So it was the period of transition that concerned me rather than the stable situation that would prevail after the transition was complete.

It was for this reason that I encouraged Stephen Barker, now a professor at Cardiff University, to turn some of his postdoctoral effort toward understanding exactly what the jumps in the ice-core electrical conductivity record were telling us. As already mentioned, Barker showed that the jumps occurred at times when a nearly perfect match existed between the inputs of acid and of calcium carbonate to the ice cap. To a chemist, the ice was close to the "titration" point. In this situation, the conductivity becomes supersensitive to slight changes in the composition of the input material. Once convinced that this was the case, I put to rest much of my remaining concern about the conveyor's potential role in ongoing global warming.

I say "remaining concern" because the consensus among ocean modelers was that only a very large warming could bring the conveyor to a halt, and that rather than occurring suddenly it would slow down gradually over the period of a century or more. If the impetus was strong enough, the decline would continue until the conveyor slowed to a halt. If it wasn't, the decline would bottom out and the conveyor would slowly recover (see figure 10-2).

In the models, the impetus for conveyor weakening comes from the addition of excess fresh water to the northern Atlantic as a result of increased rainfall and river runoff. This excess is created by the strengthening of the hydrologic cycle due to the increase in atmospheric water vapor produced by global warming. More water evaporates from the ocean and hence more falls as rain and snow.

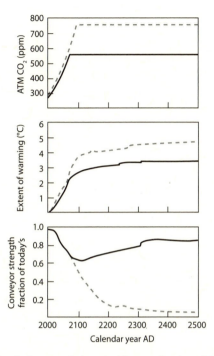

Figure 10-2. Simulations carried out by Thomas Stocker of Bern University in an ocean-atmosphere model designed to gauge the responses of global temperature and conveyor strength to long-term increases in atmospheric CO_2 content. In the scenario shown by the dark curves, the CO_2 content is increased at the rate of 1 percent per year until it reaches 560 parts per million, where it is maintained for the next four hundred years. The result is a warming of a bit more than 3°C. This warming causes the Atlantic's conveyor circulation to sag to 65 percent of its original strength; then the conveyor circulation rebounds to about 85 percent of its original strength. In the second scenario (dashed curves), the CO_2 rise continues to 760 ppm before leveling off. The result is a warming of between 4°C and 5°C, which leads to a steady drop in conveyor strength. In this case, no recovery occurs. Note that, unlike the abrupt responses of glacial time, in this simulation the shutdown is spread over the better part of two centuries.

None of the simulations invoked a catastrophic input of fresh water akin to those envisioned as the cause of the shutdowns that occurred during glacial time. The reason is that the only large reservoir of fresh water potentially vulnerable to sudden release is Greenland's ice. Although there is a consensus that as the globe warms this ice mass will slowly melt away, to date no scenario has been put forward that large parts are poised to suddenly slide into the sea and melt.

Does this mean that stoppage of the conveyor should be removed from our list of future concerns? Of course not. If we have learned anything during the past fifty years, it is that there are many aspects of the connection among the elements of the Earth's climate system that we don't fully understand. This being the case, we are surely in for surprises. Hence efforts are already under way to monitor the health of the Atlantic's conveyor circulation. What has been learned so far is that because of large seasonal and interannual fluctuation in the flow of water through the deep Atlantic, only on a multidecadal timescale will it be possible to determine whether the conveyor's operation is being perturbed by global warming. Nevertheless, the observations will continue, as will efforts by modelers. Although the conveyor should not be removed from our list of concerns, it certainly can be moved well down the list.

What about other so-called tipping points? It is currently fashionable in environmental circles to speak of irreversible thresholds that will be passed as the buildup of fossil fuel–derived CO_2 continues. Having been guilty of crying wolf, I am uncomfortable with this concept. Surely such tipping points may exist, but as we currently can only hint at what they might be, we can't predict at what level of the atmospheric CO_2 buildup they might kick in. So, as my friend Richard Alley likes to say, we are in the situation of a blind man who finds himself walking along with an inkling of suspicion that he might be headed toward an unfenced cliff.

If there are to be surprises, I suspect they will involve the hydrologic cycle. The paleo record provides an example. Scott Stine, a geomorphologist at the University of California, took advantage of a convincing paleo archive to demonstrate that, during the Medieval Warm time interval, California's Sierra Nevada mountains and the adjacent desert region experienced two century-duration droughts more severe than any of the multiyear dry spells that have occurred during the past 150 years. The archive is tree stumps now under water that Stine found in a number of locations. As the tree species of interest cannot survive extended periods of root submergence, these stumps record periods when the water was absent. Counts of growth rings pin down the duration of the drought. Radiocarbon dating fixes the time when the drought occurred. Stine's most spectacular find was eighty or so stumps in the bed of the West Walker River, which drains the west slope of the Sierra Nevada (see figure 10-3). This now sizable waterway must have been dry during the course of the two droughts. Another of Stine's locales is Lake Tenaya, a small body of water in the mountains above Yosemite Valley. Tree trunks rooted at many meters depth project just above the surface. Fed by snow melt, this lake has overflowed in all but one year since it was discovered in the late 1800s. As the trees are rooted well below the hard rock outflow spillway, not only did the lake fail to overflow during the interval of tree growth, but its inflow must have been so meager that it was lost entirely by evaporation from the surface of a closed-basin lake substantially smaller than today's.

Although less well documented, this medieval drought appears to have extended southward into Mexico. It may well account for the mysterious demise of the Anasazi culture. If what

Figure 10-3. Two views of stumps projecting above the surface of the West Walker River, which drains a portion of the east slope of California's Sierra Nevada mountains. Ring counts establish that these trees lived for more than a century before being killed when the river was reestablished after dormancy during the intense medieval drought. These stumps were discovered and photographed by Scott Stine of the University of California.

appears to have been a minor global climate perturbation could have resulted in such a major reduction in water supply, it is not difficult to imagine that the much larger climate change to be produced by the buildup of CO_2 will produce even more serious reductions in water supply.

Isaac Held of NOAA's Princeton, New Jersey, laboratory, who is one of the world's leading experts on atmospheric dynamics, makes a rather scary prediction. As the world warms, rainfall will be even more strongly focused on the tropics and, as a consequence, the extratropical dry lands will become ever more starved of moisture. Because this prediction is based on theory and backed by model simulations, it is broadly accepted.

This prediction is also consistent with evidence from the paleoclimate record. Although no adequate global analogue exists for times warmer than the present, we do have an impressive one for the last glacial maximum. If Held's prediction is correct, then during the cold glacial, the planet's dry lands should have been less arid and its tropics less moist.

The cold analogue is based on the changes in size of closed-basin lakes. These water bodies have no outlet. Rather, the water they receive from rivers and direct rainfall is lost entirely by evaporation. This being the case, their sizes fluctuate in response to changes in input and of course also to changes in evaporation. Well-known examples are the Great Salt Lake in Utah and the Dead Sea in the Middle East.

During the last glacial maximum, closed-basin lakes in the temperate dry lands were four to eight times larger than their late Holocene remnants. Many of the lakes that existed during glacial time are now playas (i.e., lakes that exist only during wet seasons). That these glacially expanded lakes were present in the western United States, in the Middle East, in northwest China, and in Argentina's Patagonian dry lands demonstrates that this is a global rather than a regional phenomenon. Although, as

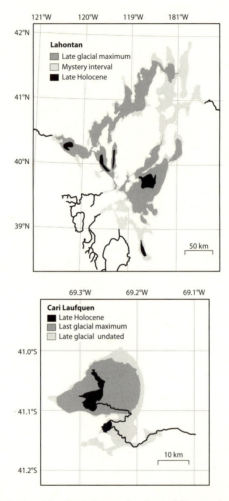

Figure 10-4. Contrast in size between the time of the maximum of the last glacial period and the late Holocene for Lake Lahontan at 40°N in the Great Basin of the United States and for Lago Cari Laufquen at 41°S in the Patagonian dry lands of Argentina. Lahontan was even larger during the Mystery Interval (17.5 to 14.5 kyrs). Cari Laufquen was also even larger at some time during the last glacial period but, despite a thorough search, no datable material has been found on the higher shorelines. Maps courtesy of Ken Adams (top) and Jay Quade (bottom).

some authors have proposed, the presence of the Laurentide ice sheet may have influenced the size of Lake Lahontan in our Great Basin,[1] and a shift in the position of the westerly winds may have influenced the size of Lago Cari Laufquen in Argentina's Patagonian dry lands, there must be some broader influence at work: namely, the reverse of Held's focus. During cold times, precipitation was less tightly squeezed into the tropics and, as a result, at least the temperate dry lands received considerably more precipitation than they do today.

An important point is the magnitude of the changes. They were very large (see figure 10-4). In order for a lake's size to increase by a factor of eight, even a doubling of the rainfall rate and a halving of the evaporation rate are not enough. Another factor of two is necessary. It is the fraction of the precipitation that runs off as opposed to evaporates from the land surface. It is a strong function of rainfall. In the eastern United States, where I live, about 50 percent of the rain that falls flows back to the sea in rivers. The rest is transpired by plants. In the dry Great Basin, the runoff fraction is only 5 to 10 percent. Rarely do the soils become sufficiently saturated with water to generate runoff. I suspect that changes in this parameter account for a sizable part of the large increase in the size of closed-basin lakes during glacial time.

What about the tropics? Were they less wet during peak glacial time? As we have seen, the monsoons were weaker during cold times and correspondingly the methane content of the atmosphere was lower. Although one would not expect to find closed-basin lakes in the wet tropics, there is one pronounced dry area: the Rift Valley of eastern Africa. It contains several large lakes. While sediment cores suggest that these lakes were

---

[1] As illustrated in figure 10-4, once large glacial-age Lake Lahontan is currently represented by four small remnants.

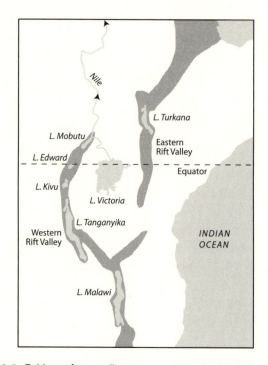

Figure 10-5. Evidence from sediment cores suggests that lakes in the African Rift Zone were smaller during glacial time than they are today. For Lake Victoria, which sits astride the equator and currently overflows into a branch of the Nile River, sediment cores studied by Tom Johnson reveal that it was once completely dry, as indicated by a buried soil horizon. This soil is covered by lake sediments whose deposition, as indicated by ¹⁴C dates, commenced at the onset of the Bølling Allerød. At the same time, Lake Lahontan underwent a major desiccation.

substantially smaller during glacial time, the most convincing evidence comes from Lake Victoria (see figure 10-5), which straddles the equator and currently overflows into a tributary to the Nile River. Four sediment cores obtained by Tom Johnson of the University of Minnesota–Duluth all encountered a soil horizon indicating that at some time in the past, this large lake not only didn't overflow but was completely dry. Moreover, radiocarbon dates on limnic material from the base of the sediments deposited on this soil horizon turn out to correspond in age to the onset of the Bølling Allerød warm interval. At that time, Lake Bonneville, the predecessor to the Great Salt Lake, and Lake Lisan, the predecessor to the Dead Sea, were undergoing major desiccations.

Although our atmosphere-ocean models and analogies to the past provide a glimpse of what is likely to come, I suspect that we are in for many surprises. As Roger Revelle so aptly predicted, we are likely to learn the true consequences only as our experiment proceeds! And as the pink and blue snake on my office wall warns, climate is an angry beast. As such, the response to extra CO_2 is unlikely to be totally predictable. I only wish that I could live long enough to experience the results of Roger's experiment. But alas, as a seventy-seven-year-old, I'll get to view only the prologue! I am happy, however, that despite the flip flops in my thinking, my idea about the great ocean conveyor turned out to be correct. Richard Alley generously titled a review article on this subject "Wally Was Right."[2]

[2] Richard Alley, "Wally Was Right: Predictive Ability of the North Atlantic 'Conveyor Belt' Hypothesis for Abrupt Climate Change," *Annual Review of Earth and Planetary Sciences* 35 (2007): 241–272.

Glossary

ablation
: The loss of mass by a glacier due to summer melting and sublimation.

abyssal
: The most remote depths of the ocean.

acid aerosols
: Tiny airborne particulates formed from acidic gases such as sulfur dioxide and nitrous oxide. The aerosols are produced when these gases react with oxygen to produce, respectively, sulfuric acid and nitric acid.

alpha particle decay
: Radioactive elements such as uranium and thorium undergo nuclear transformation by emitting helium nuclei.

Anasazi
: A sophisticated Indian culture that flourished in the American southwest and then abruptly disappeared.

anaerobic
: Waters in which there is no oxygen gas.

anion
: Atoms or molecules with a negative electric charge (i.e., electron acceptors).

antiphasing
: Cyclic changes which oppose one another.

bifurcation
: A mathematical term for situations where a system can jump from one mode of operation to another.

bioturbation
: The stirring of sediments by organisms (mainly worms).

cation
Atoms or molecules with a positive electric charge (i.e., electron donors).

convection
The overturning of an ocean water column generated when the density of waters at the surface exceeds that for the underlying water.

cosmogenic isotopes
Radioactive atoms created by the impacts of protons accelerated to extremely high velocity by magnetic fields in outer space. These protons smash the nuclei of nitrogen and oxygen atoms in our atmosphere, releasing a host of secondary particles, and the neutrons produced in this way in turn transmute atoms in air and rocks to form radioactive isotopes (^{10}Be, ^{26}Al, ^{36}Cl, etc.).

deconvolution
A term used to describe a mathematic procedure designed to reconstruct the cause from the observed result of some Earth process (for example, the air temperature histories from the down-hole temperature records in polar ice).

deglaciation
The time period when the ice sheets of full glacial time were retreating toward their interglacial size.

dendrochronology
Timescale based on tree-ring counts.

desiccation
Loss of water due to evaporation.

deuterium
A hydrogen atom with one proton and one neutron in its nucleus (i.e., ^{2}H).

diabase sill
A basaltic lava injected below ground as opposed to ejected from a volcanic vent. Because a sill cools slowly, larger crystals are formed, distinguishing it from fine-grained basalts rapidly crystallized from a lava flow.

ductile
A material that deforms like taffy rather than breaks like pretzels.

end member
One of two extremes in any series.

erratics
Large rocks left behind by melting glaciers.

firn	Partially lithified snow characterized by connected air-filled pores that allow communication with the overlying atmosphere.
geomorphic	Refers to land forms (i.e., hills, mesas, ravines, moraines, etc.).
GISP	Acronym for the American ice core drilled at Greenland's summit.
gradient	Gradual change in properties (such as elevation, temperature, pressure, latitude, etc.).
gray scale	Color gradations in sediment.
GRIP	Acronym for the European ice core drilled at Greenland's summit.
gyre	Quazi-circular flows such as those that occur in the temperate regions of the ocean.
half-life	The time required for the number of atoms of a given radioactive isotope to decrease by a factor of two.
heliomagnetic	Refers to the magnetic field generated by the ions ejected into the surroundings from the sun's dark spots.
hydrology	The study of the cycle of water.
insolation	The rain of the sun's rays on the Earth's surface.
interstadial	Millennial-duration episodes of relative warmth separating periods of extreme cold (i.e., stadials).
kyr	A millennium (one thousand years).
limnic	Concerning lakes.
lithics	Rock fragments.
meltwater	Water produced by the ablation of glaciers.
meridional	Refers to latitude.
micromole	One millionth of a mole. A mole is 6×10^{23} molecules.

Mystery Interval	The first half of the postglacial warming and CO_2 rise as recorded in Antarctic ice cores (17.5 to 14.5 calendar kyrs ago).
nanodiamonds	Microscopic diamonds invisible to the naked eye.
obliquity	The tilt of the Earth's equator with respect to its orbit.
paleoclimatology	The study of past climates.
perturbation	Small departure from the norm.
ppm	Parts per million.
precession	The cyclic wobble of the Earth's spin axis in response to the gravitational pull by the sun and moon on its equatorial bulge.
seasonality	The annual cycle of climate in response to changing solar insolation.
snowline	The boundary between those parts of a glacier undergoing net mean annual accumulation and those undergoing net mean annual ablation.
speleothems	Deposits of $CaCO_3$ formed in caves.
stadial	Millennial-duration episodes of extreme cold separating interstadial times of moderate cold.
titration point	The end point of a chemical reaction.
varves	Annually banded sediment.

GASES OF IMPORTANCE TO PALEOCLIMATE RESEARCH

H_2O	Primary greenhouse gas.
CH_4	Swamp gas.
CO_2	Important greenhouse gas.
O_2	Makes up ~20 percent of Earth's atmosphere.
N_2	Makes up ~80 percent of Earth's atmosphere.

Ions Formed by Common Chemical Elements

Ca^{++}	Calcium.
Mg^{++}	Magnesium.
K^+	Potassium.
Na^+	Sodium.
Cl^-	Chlorine.
$SO_4^=$	Sulfate.
$NO_3^=$	Nitrate.
$CO_3^=$	Carbonate.
HCO_3^-	Bicarbonate.

Radioisotopes Used in Paleoclimate Research

^{10}Be	$t\frac{1}{2} = 1.5 \times 10^6$ yrs	Produced by cosmic rays.
^{14}C	$t\frac{1}{2} = 5.7 \times 10^3$ yrs	Produced by cosmic rays.
^{87}Rb	$t\frac{1}{2} = 4.9 \times 10^{10}$ yrs	Parent of ^{87}Sr.
^{187}Sm	$t\frac{1}{2} = 1.1 \times 10^{11}$ yrs	Parent of ^{143}Nd.
^{230}Th	$t\frac{1}{2} = 7.5 \times 10^4$ yrs	Produced by the decay of ^{234}U.
^{234}U	$t\frac{1}{2} = 2.5 \times 10^5$ yrs	Produced by the decay of ^{238}U.
^{238}U	$t\frac{1}{2} = 4.5 \times 10^9$ yrs	Parent of ^{234}U.

Stable Isotopes Terminology

hydrogen ^2H, ^1H (D, H)

$$\delta D = \frac{(D/H)_{sample} - (D/H)_{std.}}{(D/H)_{std.}} \times 1000$$

carbon ^{13}C, ^{12}C

$$\delta^{13}C = \frac{(^{13}C/^{12}C)_{sample} - (^{13}C/^{12}C)_{std.}}{(^{13}C/^{12}C)_{std.}} \times 1000$$

nitrogen ^{15}N, ^{14}N

$$\delta^{15}N = \frac{^{15}N/^{14}N)_{sample} - {}^{15}N/^{14}N)_{std.}}{^{15}N/^{14}N)_{std.}} \times 1000$$

oxygen ^{18}O, ^{16}O

$$\delta^{18}O = \frac{^{18}O/^{16}O)_{sample} - {}^{18}O/^{16}O)_{std.}}{^{18}O/^{16}O)_{std.}} \times 1000$$

Figure Credits

4-1.	Original for this book; Sam Epstein, data.
4-2.	From Jim Kennett.
4-3.	From Konrad Hughen.
4-4.	From a number of labs, not named in caption.
4-5.	From George Denton
5-1.	From Gerard Bond.
5-2.	From Sidney Hemming.
5-3.	From Sidney Hemming.
5-4.	From Eric Grimm and from Helge Arz.
5-5.	From Gerard Bond.
6-1.	Original for this book.
6-2.	From Kurt Cuffy.
6-3.	From Jeff Severinghaus.
6-4.	From George Denton.
6-5.	Original for this book.
6-6.	From Broecker article.
7-1.	Original for this book.
7-2.	From Larry Edwards with permission.
7-3.	From Ed Brook.
7-4.	Original for this book.
7-5.	From Michael Bender and Jean Jouzel.
7-6.	From Jeff Severinghaus with permission.
8-1.	Original for this book.
8-2.	From Gary Comer and from Jim Teller with permission.
8-3.	From Tom Lowell with permission.
8-4.	From Stefan Rahmstorf.
8-5.	From Enrico Bonatti.
9-1.	From Richard Alley.
9-2.	From Gerard Bond.
9-3.	From Gerard Bond.
9-4.	From telescopes and satellites.
9-5.	From George Denton.
9-6.	From Lloyd Keigwin.
10-1.	From Ken Taylor.
10-2.	From Thomas Stocker.
10-3.	From Scott Stine with permission.
10-4.	Original Ken Adams and Jay Quade.
10-5.	Original for this book.

Supplementary Readings

Alley, Richard B. *The Two-Mile Time Machine: Ice Cores, Abrupt Climate Change, and Our Future.* Princeton, NJ: Princeton University Press, 2002.

Broecker, Wallace S. *The Glacial World according to Wally.* New York: Eldigio Press, 1992; second revised edition, 1995; third revised edition, 2002.

———. *The Ocean's Role in Climate Yesterday, Today and Tomorrow.* New York: Eldigio Press, 2005.

Imbrie, John, and Katherine Palmer Imbrie. *Ice Ages: Solving the Mystery.* Cambridge, MA: Harvard University Press, 1986.

Richardson, Philip L. "On the History of Meridional Overturning Circulation Schematic Diagrams." *Progress in Oceanography* 76 (2008): 466–486.

Index